AF848624

The Economics of Structural Change in Knowledge

Policymakers and researchers need to better understand how knowledge creates economic and social value. They can only do this if they have better theoretical and analytical tools. *The Economics of Structural Change in Knowledge* provides new help on both these levels.

David Rooney, *University of Queensland Business School, Australia*

This book provides a wonderful introduction to the role that information and knowledge play in the determination of economic growth and fluctuations in modern economies. Its synthesis of Schumpeterian and Kuznetsian traditions into a clear, integral framework is much needed and will be of value to students and researchers alike.

Dennis Patrick Leyden, *The University of North Carolina at Greensboro, USA*

This book provides an elaboration upon the concept of knowledge from an economic viewpoint. However this is not a book on economics of knowledge, at least not in the conventional sense. Most of the existing books on the matter have focused on the treatment of knowledge in terms of properties of knowledge as an economic good, incentive schemes for the creation of knowledge, issues about the codified/tacit nature of knowledge and the like.

This book attempts to develop a different approach, by adopting a systemic perspective to knowledge itself, as well as to its generation, diffusion and utilization. A systemic calls for the conceptualization of knowledge as a sort of "tangible" (although, as is well known, it is an intangible asset) endowed with a structure of its own, the same way as a laptop or a smartphone are systems made of interacting components organized in specific ways. Knowledge is regarded in this book as structured item, and structural change refers here to the changes that the architecture of such structure can undergo over time. The argument is built by moving from the appreciation of the most traditional uses of the term "structural change" of the concept of knowledge in economics.

By reviving a structuralist approach to economics, and emphasizing the dynamic aspects, we articulate a complexity-based approach able to combine the view of structural change in knowledge with the traditional approach to structural change in economics, and eventually pave the way to a much comprehensive understanding of the dynamics governing the functioning of socio-economic systems. The empirical investigations at different levels of aggregation presented in the third part of the book confirm the usefulness of the theoretical and methodological framework both for the purposes of analysis and for the elaboration of a new approach to technology policy.

Francesco Quatraro is Associate Professor at the University of Nice, GREDEG-CNRS and Research Associate at the BRICK, Collegio Carlo Alberto, Torino, Italy.

Routledge Studies in Global Competition
Edited by John Cantwell, Rutgers
The State University of New Jersey, USA
and
David Mowery
University of California, Berkeley, USA

1 **Japanese Firms in Europe**
Edited by Frédérique Sachwald

2 **Technological Innovation, Multinational Corporations and New International Competitiveness**
The case of intermediate countries
Edited by José Molero

3 **Global Competition and the Labour Market**
Nigel Driffield

4 **The Source of Capital Goods Innovation**
The role of user firms in Japan and Korea
Kong-Rae Lee

5 **Climates of Global Competition**
Maria Bengtsson

6 **Multinational Enterprises and Technological Spillovers**
Tommaso Perez

7 **Governance of International Strategic Alliances**
Technology and transaction costs
Joanne E. Oxley

8 **Strategy in Emerging Markets**
Telecommunications establishments in Europe
Anders Pehrsson

9 **Going Multinational**
The Korean experience of direct investment
Edited by Frédérique Sachwald

10 **Multinational Firms and Impacts on Employment, Trade and Technology**
New perspectives for a new century
Edited by Robert E. Lipsey and Jean-Louis Mucchielli

11 **Multinational Firms**
The global–local dilemma
Edited by John H. Dunning and Jean-Louis Mucchielli

12 **MIT and the Rise of Entrepreneurial Science**
Henry Etzkowitz

13 **Technological Resources and the Logic of Corporate Diversification**
Brian Silverman

14 **The Economics of Innovation, New Technologies and Structural Change**
Cristiano Antonelli

15 European Union Direct Investment in China
Characteristics, challenges and perspectives
Daniel Van Den Bulcke, Haiyan Zhang and Maria do Céu Esteves

16 Biotechnology in Comparative Perspective
Edited by Gerhard Fuchs

17 Technological Change and Economic Performance
Albert L. Link and Donald S. Siegel

18 Multinational Corporations and European Regional Systems of Innovation
John Cantwell and Simona Iammarino

19 Knowledge and Innovation in Regional Industry
An entrepreneurial coalition
Roel Rutten

20 Local Industrial Clusters
Existence, emergence and evolution
Thomas Brenner

21 The Emerging Industrial Structure of the Wider Europe
Edited by Francis McGowen, Slavo Radosevic and Nick Von Tunzelmann

22 Entrepreneurship
A new perspective
Thomas Grebel

23 Evaluating Public Research Institutions
The U.S. Advanced Technology Program's intramural research initiative
Albert N. Link and John T. Scott

24 Location and Competition
Edited by Steven Brakman and *Harry Garretsen*

25 Entrepreneurship and Dynamics in the Knowledge Economy
Edited by Charlie Karlsson, Börje Johansson and Roger R. Stough

26 Evolution and Design of Institutions
Edited by Christian Schubert and Georg von Wangenheim

27 The Changing Economic Geography of Globalization
Reinventing space
Edited by Giovanna Vertova

28 Economics of the Firm
Analysis, evolution and history
Edited by Michael Dietrich,

29 Innovation, Technology and Hypercompetition
Hans Gottinger

30 Mergers and Acquisitions in Asia
A global perspective
Roger Y. W. Tang and Ali M. Metwalli

31 Competitiveness of New Industries
Institutional framework and learning in information technology in Japan, the U.S. and Germany
Edited by Cornelia Storz and Andreas Moerke

32 Entry and Post-Entry Performance of Newborn Firms
Marco Vivarelli

33 **Changes in Regional Firm Founding Activities**
A theoretical explanation and empirical evidence
Dirk Fornahl

34 **Risk Appraisal and Venture Capital in High Technology New Ventures**
Gavin C. Reid and Julia A. Smith

35 **Competing for Knowledge**
Creating, Connecting and Growing
Robert Huggins and Hiro Izushi

36 **Corporate Governance, Finance and the Technological Advantage of Nations**
Andrew Tylecote and Francesca Visintin

37 **Dynamic Capabilities Between Firm Organisation and Local Systems of Production**
Edited by Riccardo Leoncini and Sandro Montresor

38 **Localised Technological Change**
Towards the economics of complexity
Cristiano Antonelli

39 **Knowledge Economies**
Innovation, organization and location
Wilfred Dolfsma

40 **Governance and Innovation**
Maria Brouwer

41 **Public Policy for Regional Development**
Edited by Jorge Martinez-Vazquez and François Vaillancourt

42 **Evolutionary Economic Geography**
Location of production and the European Union
Miroslav Jovanovic

43 **Broadband Economics**
Lessons from Japan
Takanori Ida

44 **Targeting Regional Economic Development**
Edited by Stephan J. Goetz, Steven C. Deller and Thomas R. Harris

45 **Innovation, Knowledge and Power**
Theodora Asimakou

46 **Creativity, Innovation and the Cultural Economy**
Edited by Andy C. Pratt and Paul Jeffcutt

47 **Coopetition Strategy**
Giovanni Battista Dagnino and Elena Rocco

48 **Knowledge Intensive Entrepreneurship and Innovation Systems**
Evidence from Europe
Edited by Franco Malerba

49 **Innovation in Complex Social Systems**
Edited by Petra Ahrweiler

50 **Internationalization, Technological Change and the Theory of the Firm**
Edited by Nicola De Liso and Riccardo Leoncini

51 **Territory, Specialization and Globalization in European Manufacturing**
Helena Marques and Francisco Puig

52 **Institutional Diversity and Innovation**
Continuing and Emerging Patterns in Japan and China
Cornelia Storz and Sebastian Schäfer

53 **Innovation and Economic Crisis**
Daniele Archibugi and Andrea Filippetti

54 **The Communications Industries in the Era of Convergence**
Catherine Mulligan

55 **Innovation, Technology and Knowledge**
Charlie Karlsson, Borje Johansson and Roger R. Stough

56 **Evolution of Competition Laws and their Enforcement**
Pradeep S. Mehta

57 **The Economics of Structural Change in Knowledge**
Francesco Quatraro

The Economics of Structural Change in Knowledge

Francesco Quatraro

Routledge
Taylor & Francis Group
LONDON AND NEW YORK

First published 2012
by Routledge
2 Park Square, Milton Park, Abingdon, Oxon OX14 4RN

Simultaneously published in the USA and Canada
by Routledge
711 Third Avenue, New York, NY 10017

Routledge is an imprint of the Taylor & Francis Group, an informa business

© 2012 Francesco Quatraro

The right of Francesco Quatraro to be identified as author of this work has been asserted by him in accordance with the Copyright, Designs and Patent Act 1988.

All rights reserved. No part of this book may be reprinted or reproduced or utilized in any form or by any electronic, mechanical, or other means, now known or hereafter invented, including photocopying and recording, or in any information storage or retrieval system, without permission in writing from the publishers.

Trademark notice: Product or corporate names may be trademarks or registered trademarks, and are used only for identification and explanation without intent to infringe.

British Library Cataloguing in Publication Data
A catalogue record for this book is available from the British Library

Library of Congress Cataloging in Publication Data
Quatraro, Francesco.
The economics of structural change in knowledge / Francesco Quatraro.
p. cm.
1. Knowledge management. 2. Knowledge management–Economic aspects. 3. Economics. 4. Technological innovation–Economic aspects.
I. Title.
HD30.2.Q38 2011
338.5–dc23 2011040236

ISBN: 978-0-415-56543-1 (hbk)
ISBN: 978-0-203-12327-0 (ebk)

Typeset in Times New Roman
by Wearset Ltd, Boldon, Tyne and Wear

Contents

Figures

Tables

Foreword

The manner in which wealth is created from knowledge remains one of the great enduring puzzles of economic life. So much is known in the modern world, and yet nations, regions and firms are marked by great unevenness in their practical access to knowledge. Knowledge and understanding can grow with combinatorial speed but real economies grow exponentially. How is this so? The question has occupied serious students of economic progress since Adam Smith formulated the thesis that a division of labour is a division of knowing. Marx, Marshall and Schumpeter have taken up this broad theme and found the dynamics of capitalism to reside in the ways that it develops and uses gains in human understanding. It is clearly a very uneven process and there seems little merit from their perspective in the idea of a proportionally expanding economy, an expanding stationary state in which all activities grow as one. This book captures the problem at its core by pointing to the fact that growth never occurs without a corresponding development, the latter induces the former. So growth always engages with structural change at multiple levels of an economy and we are led to enquire why the growth rates of distinct activities should be different. Kuznets, Burns and others led the way but their deep insights were lost as economic growth and development theory went macro, post-1945, and so swept under the carpet the processes by which knowledge and growth are intertwined. Quatraro's book is a timely reminder of a different way in which growth can be studied, one that puts the uneven growth of knowledge at its centre. By drawing on the retardation theorists of the 1930s, he renders comprehensible the manifest features of structural change and the vast diversity of rates of growth and profitability found in any economy at a particular point in time. By further linking development to the structure of networks of knowing he is able to connect the development of ICTs and biotechnology to the underlying growth of knowledge and to do the same for Italian and European regions as his exemplars. By treating knowledge as a structure that is contained within a structure of economic life we are pointed to a world that cannot be in equilibrium, nor can it be chaotic. Instead it is a highly ordered world that generates its own challenges to the status quo from within. Quatraro is writing in the grand tradition of magnificent dynamics, in which capitalism is rendered an open system by the nature of its dependence on knowledge. Like all open systems its future is not knowable that is the paradox on which our economic world rests.

Stan Metcalfe
Manchester
3 November 2011

Acknowledgements

This book is the outcome of the last three years of my research activity. Most of the research activity has been conducted while at the GREDEG (Groupe de Recherche en Droit, Economie et Gestion), a joint research department of French CNRS and University of Nice Sophia Antipolis. My permanence at the BRICK, Collegio Carlo Alberto and the Deparment of Economics of the University of Torino has also been especially fruitful. I owe the greatest intellectual debt to Cristiano Antonelli, Stan Metcalfe and Pier Paolo Saviotti. Without their stimuli, discussions and comments on the topics treated in this work, this book could not have been possible.

I also wish to thank the people with whom I have had the chance to interact, and who have provided indirect and direct support to this initiative. My special thanks for reading parts of the text and providing useful comments and suggestions go to: Ron Boschma, Davide Consoli, Koen Frenken, Lionel Nesta, Pier Paolo Patrucco, Dimitri Pontikakis, Andrè Rodriguez-Pose and Allen J. Scott. I'm particularly obliged to Jackie Krafft, colleague and co-author, who has coordinated most of the research projects that have contributed to the realization of the research activity.

I acknowledge the financial support of the Agence Nationale de la Recherche (ANR) through the project ANR JCJC06_141306, "Knowledge Intensive Sectors: Models and Evidence"; the Provence-Alps-Cote d'Azur region through the project "Knowledge Intensive Sectors: Models and Evidence" (Programme Volet Général); the European Union Directorate for Research, with the Grant number 266959 to the research project "Policy Incentives for the Creation of Knowledge: Methods and Evidence" (PICK-ME), within the context Cooperation Program/Theme 8/Socio-economic Sciences and Humanities (SSH) and the Collegio Carlo Alberto with the project IPER.

Last, and most, I wish to thank Alessandra for her support and patience throughout the realization of this volume, even when the weight of the baby in her womb had become particularly challenging. This book is dedicated to Riccardo, whose birth switched on another wonderful light in our lives.

Part I

Overview

1 Structural change and the knowledge-based economy

An international overview

1.1 Introduction

The economies of advanced capitalist countries have been experiencing the process of a dramatic reshaping of their structure for some decades. To be fair, such a process of structural change has consistently interested economic systems, as it was manifested formerly as the shift from agriculture to non-agricultural activities, and latterly as the movement from manufacturing sectors to service activities. In other words, structural change is an inherent characteristic of capitalist economies, and it is both a cause and a determinant of restless economic growth (Metcalfe, 2002). While this phenomenon attracted the attention of leading economists in the past, there is scant attention paid today to the economic analysis of structural change and to its relationships with other key dynamics like technological change and economic development.

This chapter aims at providing evidence of the empirical relevance of structural change in the present economic conditions. While by structural change one can mean different things, our descriptive effort will be based on the most traditional usage of the term, which refers to the change in the sectoral composition of modern economies. We will then provide empirical evidence of the links between the way structural change actually takes places in advanced countries (i.e. the increasing weight of service activities), the dynamics of technological change, with particular respect to the creation, diffusion and exploitation of information and communication technologies (ICTs), and the increasing centrality of knowledge exchanges within production processes. The emerging picture will represent a sort of empirical context to frame the analysis conducted in the rest of the book, which aims at extending the application of a structuralist approach to the analysis of technological knowledge and of the networks of knowledge-generating agents, by showing that these are strictly intertwined and that they are tied by a set of recursive feedbacks and loops such that they can be effectively accommodated by using the heuristic tools of complexity theory.

This chapter is organized as follows. The next section will provide a synthetic overview on the different theoretical contexts within which the term "structural change" is actually used in the field of economics. Section 1.3 will discuss data on the evolution of value added and employment share in manufacturing and service activities across advanced countries, so as to show the topicality of the analysis of structural change. In Section 1.4 we stress how the transition towards service-based economies made possible a change in the technological paradigm leading to the creation and diffusion of ICTs. A mutually enforcing dynamics between structural and technological change can be devised in this respect. Section 1.5 puts forward some key implications of structural change and ICT diffusion, i.e. the increasing relevance of knowledge utilization for production purposes. Data on knowledge production will show how faster rates of growth can be observed in correspondence of relatively higher shares of service activities with respect to manufacturing ones. Section 1.6 will draw some preliminary conclusions, by emphasizing the need to rejuvenate the study of structural change by extending its domain of application and integrating perspectives on different aspects of economic activities.

1.2 On the different meanings of structural change

The term "structural change" is far from having a unequivocal meaning in the field of economics. For example, in econometric theory, the issue of structural change refers to the behaviour of the parameters of a model in the course of time. The usual assumption of stationarity is commonly made, according to which one or all of the relevant parameters of the econometric model are constant over time. However, *structural breaks* can occur, such that one of these parameters changes at some time in the sampled period. The econometrics of structural change allows the identification of structural breaks in a time series by providing a rich set of tests (Hansen, 2001).

On a different ground, the concept of economic structure plays an important role in the field of industrial organization, with particular respect to the structure–conduct–performance paradigm. In this approach, the economic performance of an industry is a function of the behaviour of buyers and sellers which, in turn, is a function of the industry's structure (Bain, 1956). Industry structure includes here some variables like the number and size of economic agents, technology, barriers to entry, the extent of vertical integration and the degree of product differentiation (Scherer, 1980; McWilliams and Smart, 1993).

Within the localized technological change approach, the term structural change is used to indicate changes in relative prices of production factors. The change in relative factors, in contexts characterized by high irreversibility and bounded rationality, is likely to engender a reaction in economic agents that are confronted with two alternatives, i.e. either adapt or innovate. Both of these alternatives imply some costs for economic agents, i.e. switching costs or innovation costs. When switching costs are relatively higher due to irreversibility of previous production choices, innovating may turn out to be a better solution. The introduction of technological change appears therefore as an outcome of pressures coming from the changing conditions of factor markets, and it is directed towards the increasing exploitation of the production factor which has become cheaper. These dynamics in turn are likely to introduce further alterations of the economic structure, engendering further innovation efforts (Antonelli, 2003).

However, in a more traditional perspective, the notion of structural change is related to changes in the patterns of sectoral composition of countries and regions over the process of economic development. The pillars of this line of inquiry are usually found in the seminal works by Simon Kuznets (1930) and Arthur Burns (1934). Their works indeed provide early and impressive empirical evidence concerning the rise, growth and fall of industrial sectors and the linked shift in the main sources of industrial leadership in different countries. The economic development of countries and regions is, in this perspective, strictly tied to the performance of their leading industries, and the ability to maintain an enduring competitive advantage is strongly influenced by the ability to foster the establishment of industries in the growing phase of their development.

A much overlooked influence on this strand of analysis comes from the somewhat less celebrated work *Industry and Trade* by Alfred Marshall (1919). In this book the key factors underlying the trade between nations are analysed, by emphasizing the cyclical behaviour of industry performances and the evident relationships between a country's industrial specialization and its economic leadership. The rise and fall of British economic power are analysed in this perspective, and contrasted with the emergence of German and French industrial leadership. Marshall also emphasized the importance of production techniques in shaping a country's competitive advantage, as well as the availability of innovative inputs to the production process. Moreover, Marshall stresses throughout the book that considerations about the dynamics of industry and trade among countries can be very easily adapted to the analysis of economic interactions among regions or even smaller territorial units. In this sense, he provides a much wider toolkit to understand secular changes within economic systems, by anticipating not only Kuznets' and Burns' speculations, but also touching on some key issues that would be further developed by Joseph Schumpeter (1939, 1942) and François Perroux (1955).

In this volume we will focus on the process of structural change as conceived in this last and most influential strand of analysis. Although the origins of this approach date back to about a century ago, we will show in the next section the relevance of investigations in this field in the present economic conditions, and maintain in the rest of the chapter that the cross-fertilization with economics of knowledge is necessary in a context shaped by the transition towards the knowledge-based economy.

1.3 The timeliness of the analysis of structural change in modern economies

The effects of structural change on the process of economic development have recently received renewed attention. On the one hand, some studies deal with structural change by focusing on the consequences of both the changing specialization of national economies in favour of "hi-tech" activities and the gap with countries specializing in "low-tech" activities (Fagerberg, 1994, 2000). On the other hand, some authors investigate the effects of structural change on the returns to R&D activity and on the tendency of the rate of profit to fall (Frantzen, 2000; Wolff, 2003; Quatraro, 2009a).

The topicality of structural change is well reflected in the data on the changing distribution of employees across industries in the different advanced countries. While in the 1970s much of the attention of economists was given to the shift of employment from agriculture to manufacturing activities, in recent years a gradual shift from manufacturing to the service sector has begun to characterize the dynamics of advanced capitalist economies.

The data reported in Figure 1.1 provide an impressive sketch of such a process. The share of employees in four sectors – manufacturing, financial

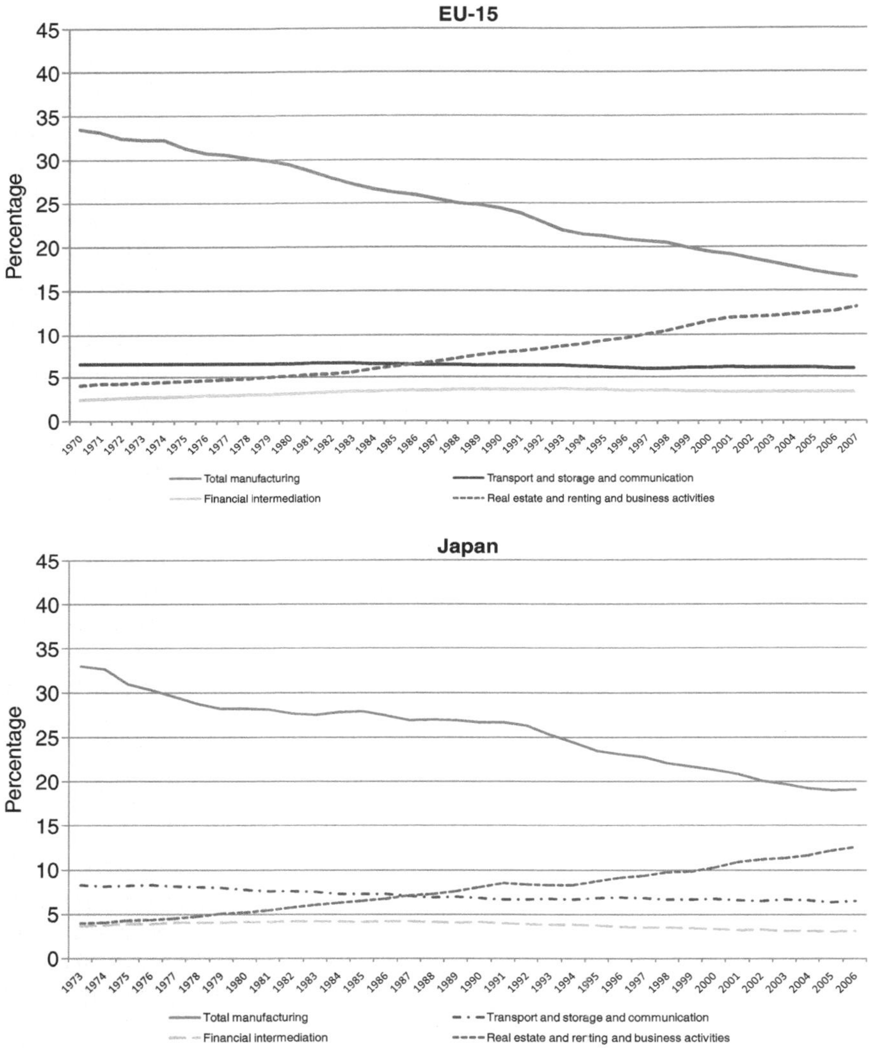

Figure 1.1 The evolution of manufacturing share of employment across EU-15, US, Japan and Korea.

intermediation, transport and storage and communication, real estate and renting and business activities – have been calculated on the basis of data supplied by the Groningen Growth and Development Centre.[1] The diagram compares the four main economies. The evidence about the EU-15 aggregate and the US looks very similar. Even though we lack US data for the first half of the 1970s, we can see that for both regions the share of employees in manufacturing falls over the whole observed period. However, in the late 1990s the rate of fall in the US is slightly faster than the EU-15 so that at the end of the observed period, i.e. in

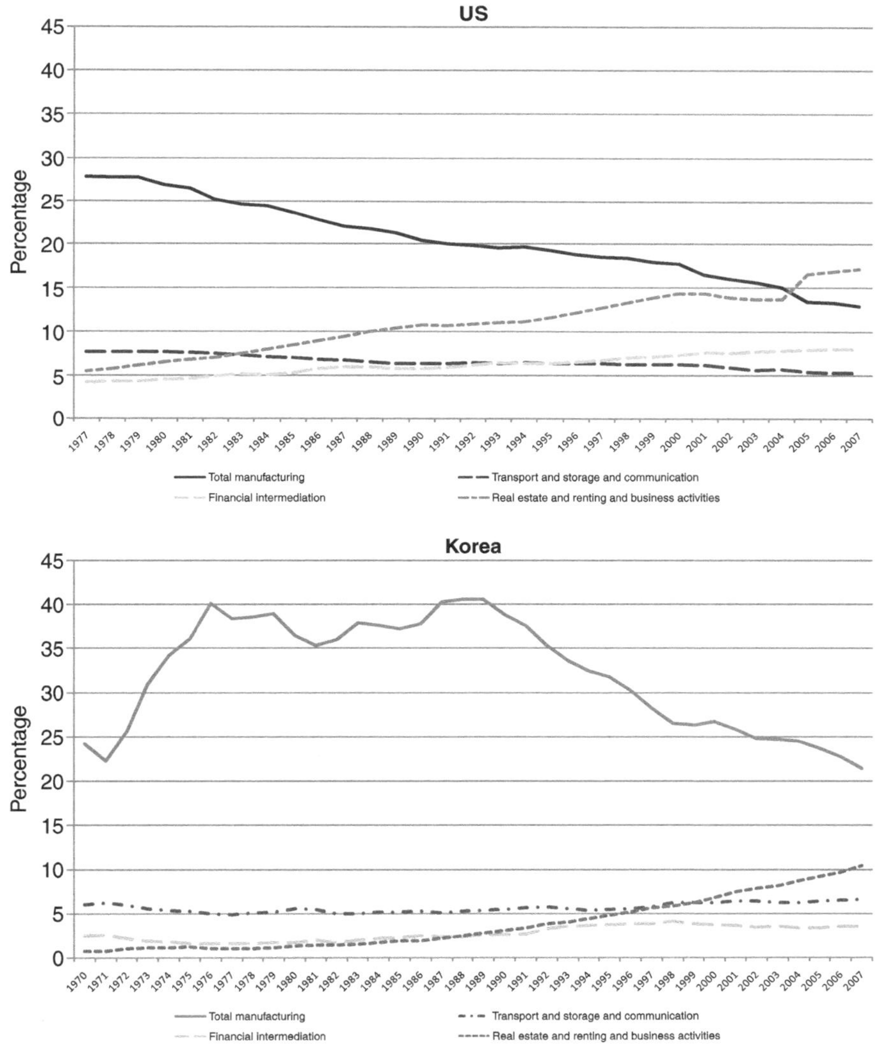

Figure 1.1 continued.

2007, the share is about 13 per cent in the US and 16 per cent in the EU-15. The dynamics of Japan slightly drift from such observed trend. While in Japan the initial share of manufacturing employees was about 33 per cent, like in the EU, we observe a steep decrease up to 1975, and then a prevalent stationary dynamics around 27 per cent just until 1990. Then we observe a modest decrease until 2000, such that the share never falls below 20 per cent. Interestingly enough, since 2001 the manufacturing employment share started increasing, though at a pretty slow rate. A very different situation can be observed in the case of Korea. In the early 1970s such employment share was just below 25 per cent, but it started immediately to increase at a very fast rate until 1976, so that it had arrived at 40 per cent within five years. Then it remained rather stable until 1988 when it started to decrease, eventually dropping to about 21 per cent in 2007. It seems as if Korea went through the two different aggregate phases of strengthening and subsequent decrease of manufacturing activities in a relatively short time span.

We can reasonably conclude by now that a process of gradual decrease in the weight of manufacturing activities is ongoing in the four key observed economies. It is as reasonable to wonder where employees leaving the manufacturing sectors are directed. Figure 1.1 shows interesting dynamics concerning the sector involving real estate and renting and business activities. The most relevant evidence pertains to the US. Indeed one can observe that the employment share in such service activities continuously increases over time at such a pace that in 2004 they outperformed manufacturing activities and attained some 17 per cent share of total employment. The US evidence is thus very informative and can be considered as archetypal of the much debated transition process to the service economy, which, after all, is exactly the effect of a dynamics of structural change. The EU-15 performance is not comparable in terms of magnitude, although it has followed the same trend. The share of private business services indeed began to grow in the early 1980s, while in the 1970s in the US it was already increasing, but it remained well below manufacturing over the observed time span, so that in 2007 it was about 14 per cent. The Japanese evidence looks very interesting in this respect. Indeed, although the employment manufacturing share therein remained around 20 per cent in the 2000s, real estate and renting and business activities grew so much that in 2007 they also approached the 20 per cent share, which is far above the US and the European evidence. Therefore, Japan seemed to have caught up and actually outperformed both US and Europe in this respect.

It can be useful now to look at within-Europe differences in terms of evolutionary patterns of employment in manufacturing and service sectors. Figure 1.2 shows the evidence concerning Germany, France, Italy and Spain, which may be taken as representative of most advanced continental European countries. As expected, the country showing the highest share of manufacturing employment in the 1970s is Germany (about 40 per cent), followed by Italy (about 35 per cent) and then Spain and France on similar values (about 30 per cent). The differential dynamics are very interesting. The country that appeared to have most keenly pursued the transition from manufacturing to a service-based economy is France. The manufacturing employment share fell from 30 per cent in 1970 to 14 per cent in 2007, while real estate and renting and business activities increased to about 16 per cent. The French data very much resembles that of the US. The German evidence is also characterized by a marked decrease of the manufacturing share, which in 2007 was half the value of 1970, as in France. The same also applies to Spain. In Italy the situation is slightly different, as the decreasing trend emerged relatively late. On the contrary, in the first half of the 1970s the manufacturing share showed a slight increase and then remained stable up to the end of the decade. One can observe a fall in the manufacturing share of employment in Italy only in the early 1980s, and at such a rate that in 2007 it was still about 23 per cent, i.e. even higher than the German situation. This suggests a relative delay in Italy compared to the other advanced European countries.

Figure 1.3 shows data concerning some north European countries: the UK, Sweden, Denmark and Finland. The UK data is the most evident European case of transition towards a service-based economy. The share of manufacturing employment falls consistently over the observed period from about 35 per cent in 1970 to about 10 per cent in 2007, i.e. a fall of about 71 per cent, with an average annual growth rate of −1.8 per cent. On the other hand, the share of real estate and renting and business activities shows an enduring increase, which is much more marked throughout the 1980s. Interestingly enough, the financial intermediation sector is also characterized by significant growth, at the turn of the decade between the 1970s and the 1980s. The three remaining countries share a common dynamics of manufacturing employment share, which is at about 30 per cent in 1970 and consistently falls during the observed period. Such a share remains, however, well above the 15 per cent in the case of Sweden and Finland in 2007, while it reaches 14 per cent in Denmark. Concerning the dynamics of service sectors, and in particular of the real estate and renting and business activities, we can observe a common increasing trend, though at evidently different rates. In Sweden the service employment share increases at a rate that demonstrates a marked acceleration in the second half of the 1980s, while Denmark experiences such a boost at the end of the 1990s and the beginning of 2000s. The situation is a bit different in Finland, where the increase of service sectors is relatively smooth.

The gradual shift from a manufacturing-centric to a service-based economy does not involve only the reallocation of employees across industries, which is a somewhat long-lasting process, but also and mainly a change in the locus of value creation. Figure 1.4 shows the evolution of value-added share in the four sectors across the EU-15, US, Japan and Korea. The data about the value-added share are of course in line with the evidence concerning the evolution of employment share. However, we can notice how the dynamics of value added seems to anticipate that of employment. In the case of the US, for example, the value-added share of real estate and renting and business activities overtake that of manufacturing sectors already in 1986, while the same occurs in the employment share only in 2004. The same evidence can also be observed in the case of EU-15 countries. The value-added share of private business services overtakes that of manufacturing in the early 1990s, while there is no evidence yet of such overtaking regarding the employment data. We could reasonably expect to observe it in a few years.

If we move to continental European countries (Figure 1.5), the impression of a lag between value added and employment dynamics receives further support. The reallocation of employment across sectors is likely to follow the reallocation of

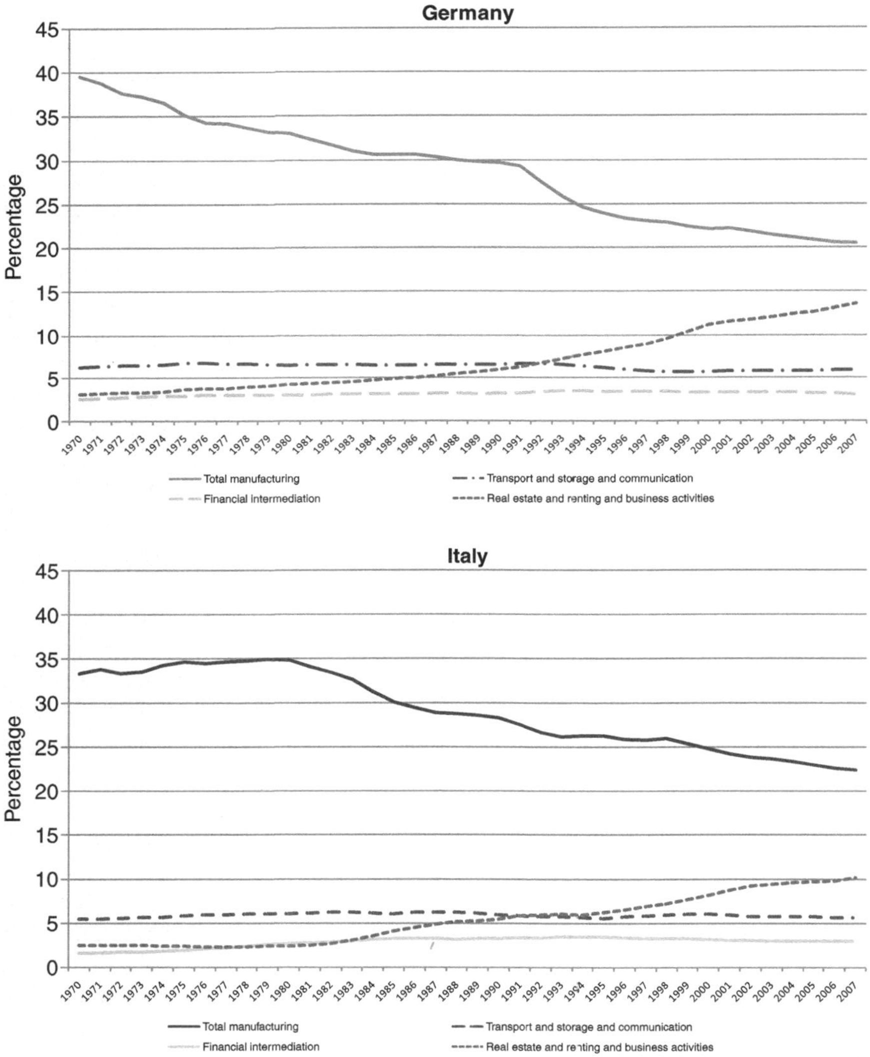

Figure 1.2 The evolution of manufacturing share of employment across Germany, France, Italy and Spain.

value added. This is fairly evident in the case of France, where we can see that the intersection between manufacturing and private business sectors occurs in 1986 concerning value added and in 2002 concerning employment. In the cases of Germany, Italy and Spain, such overtaking of services' employment share is never observed, while in the case of value added it is observed in 1998, 2001 and 2004 respectively. This would suggest that even in continental Europe the transition to a service-based economy is about to gain momentum, although with a significant delay with respect to the US. Finally, Figure 1.6 shows the evidence

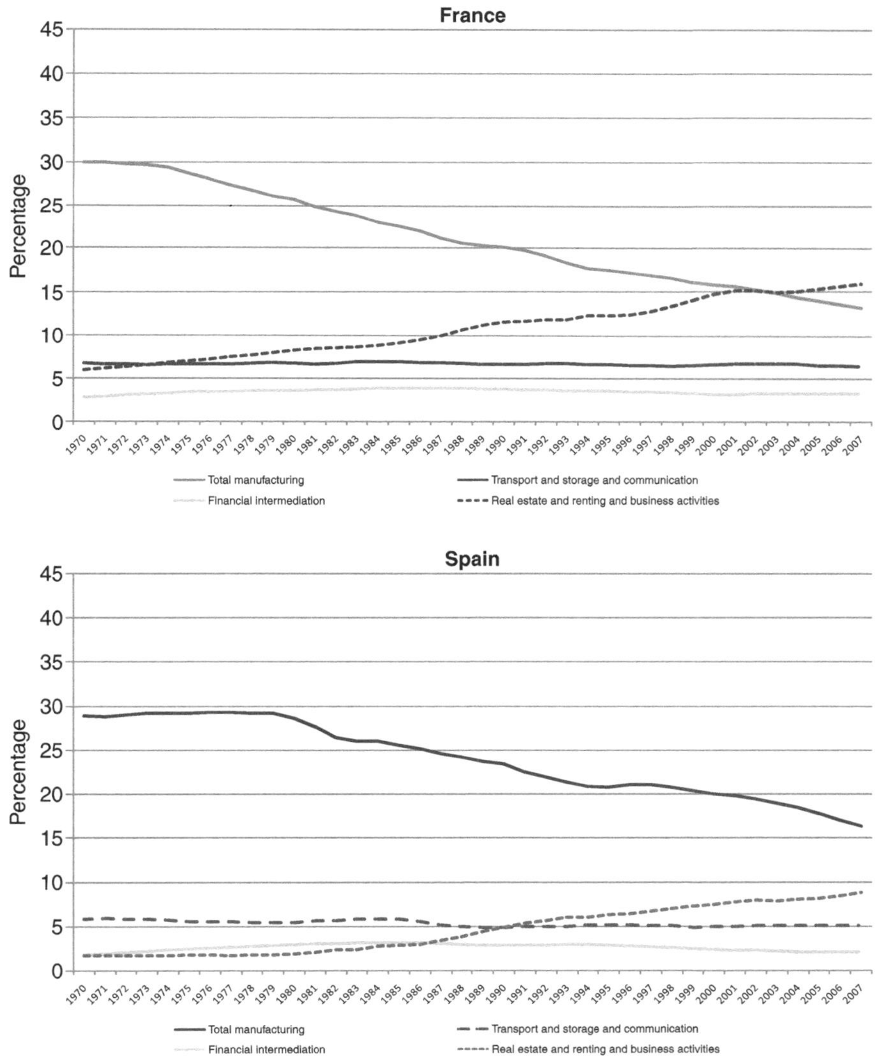

Figure 1.2 continued.

regarding the value-added share in North European countries. The main trend is also confirmed by these data, though the UK case suggests a shorter lag between the overtaking of valued-added share and that of employment. It is interesting to note how the increasing weight of service activities appears to be well established in the 2000s in Denmark and Sweden according to value-added data, which also suggests the converging evidence of employment share in a decade.

The evidence presented so far speaks for the topicality of the analysis of structural change. Although such a line of enquiry dates back to very remote

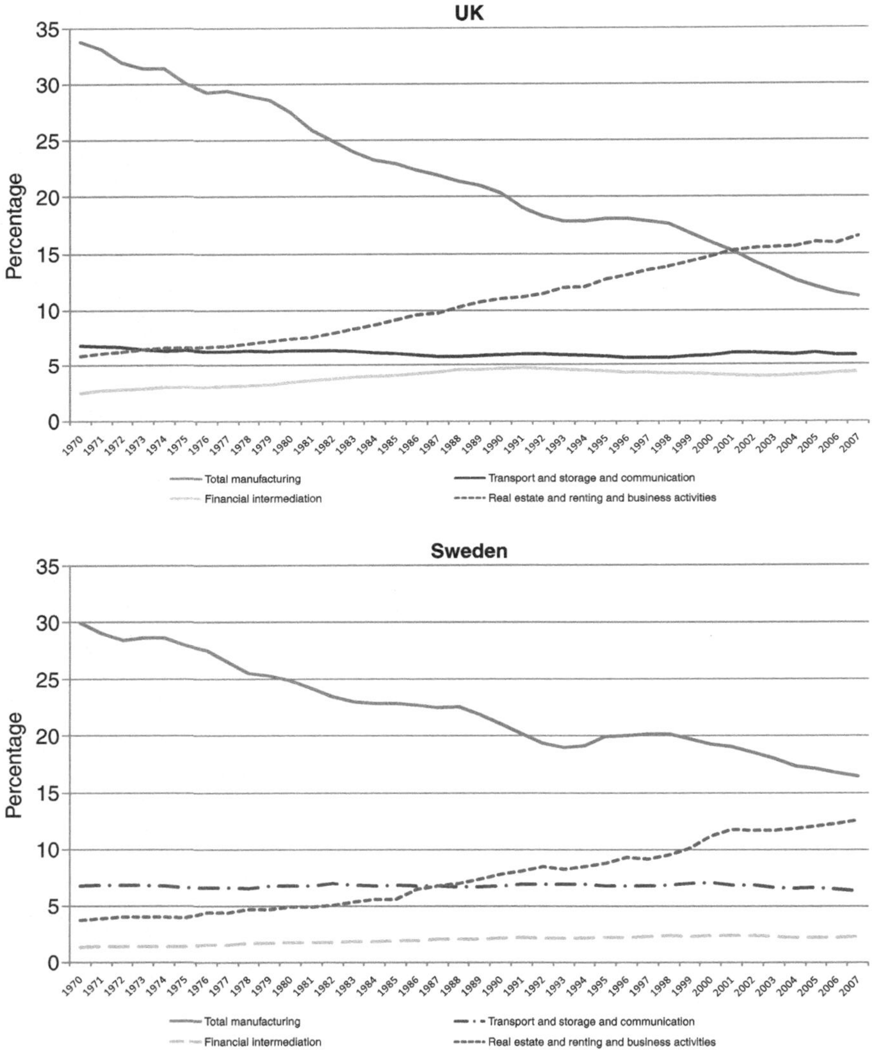

Figure 1.3 The evolution of manufacturing share of employment across UK, Sweden, Finland and Denmark.

times, the structure of advanced capitalist economies tends, by a continuous pressure, to development and mutation. Structural change is an intrinsic characteristic to the process of restless economic growth, of which it is both a cause and a consequence. It is important to stress that structural change involves the industrial composition of economic systems, but it is not limited to this. It implies indeed a multidimensional concept which is related also to changes in the size distribution of firms and in the organization of production activities. More recently, the evidence concerning the productivity surge of the US in the

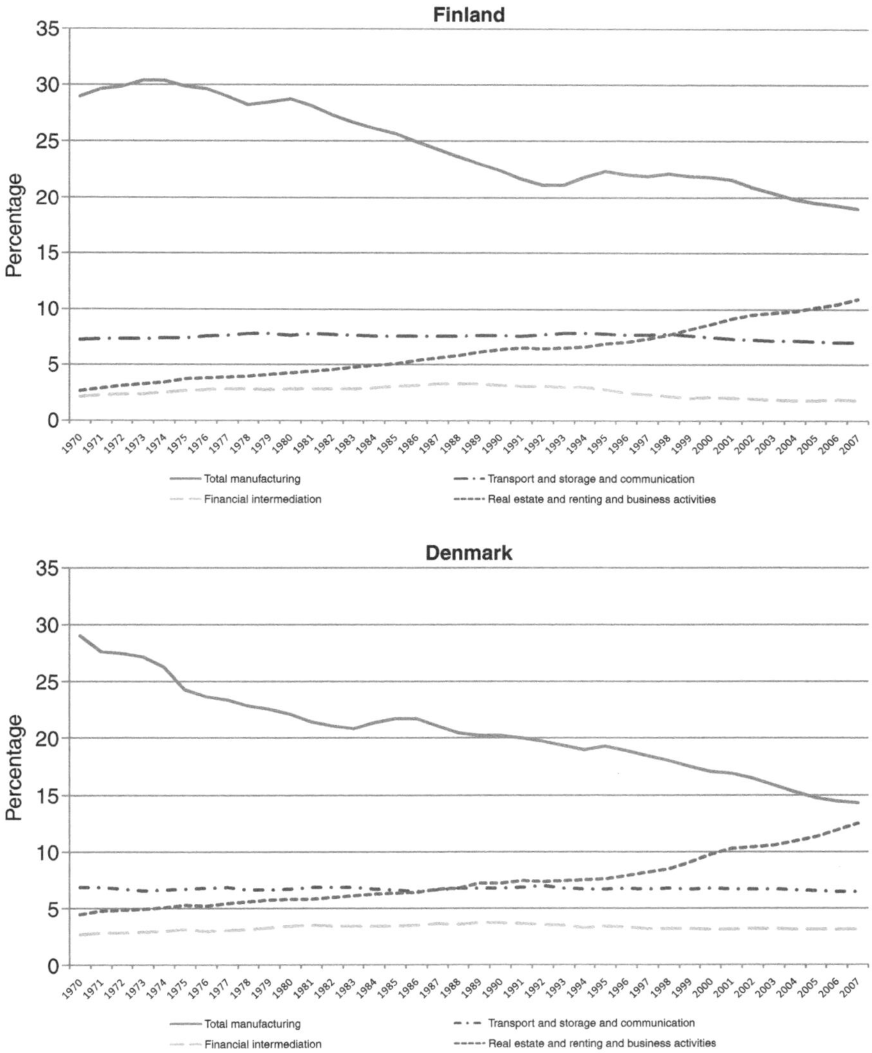

Figure 1.3 continued.

1990s stimulated a debate on its causes, which introduced another piece to the puzzle of structural change, i.e. the contribution of information and communication technologies (ICTs) to the development and establishment of service sectors. The next section will discuss such an aspect of the process, and provide the basis to introduce another key element, which concerns the increasing relevance of knowledge both as an input and as an output in advanced capitalist economies.

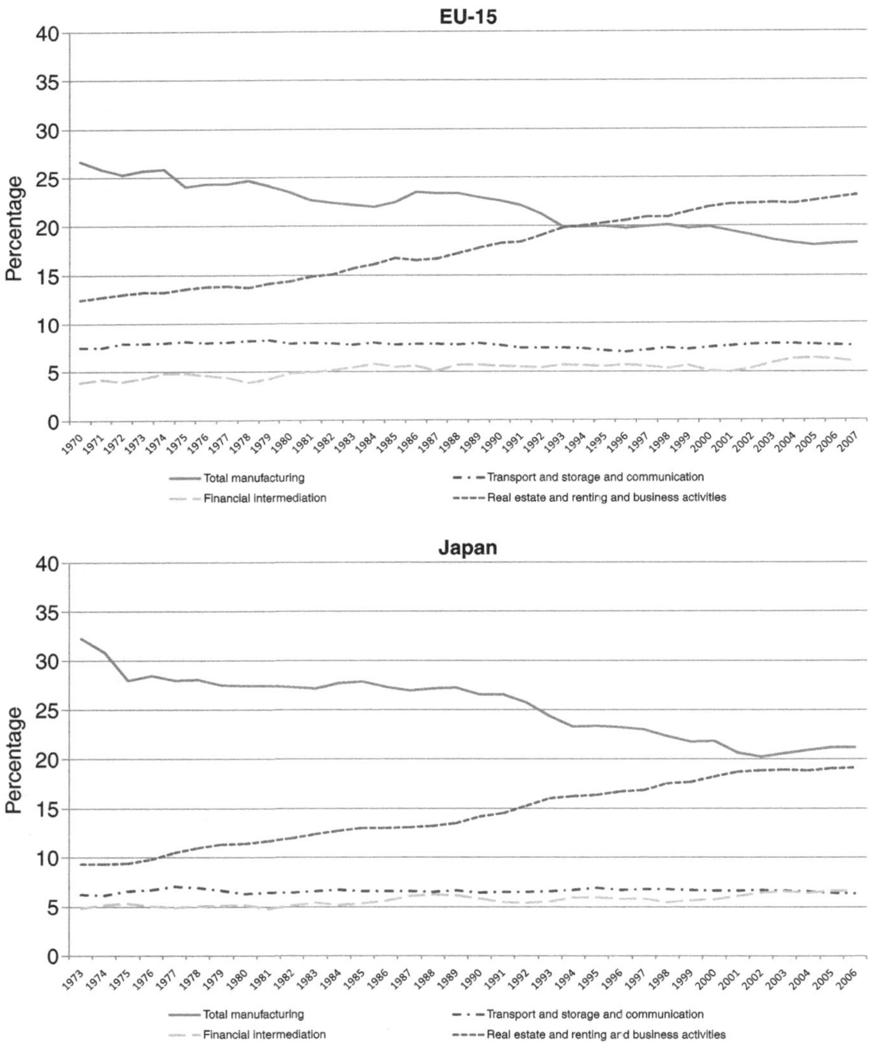

Figure 1.4 Evolution of value added share across EU-15, US, Japan and Korea.

1.4 The role of ICTs in the recent dynamics of structural change

The changing composition of industrial activities, with particular respect to the increasing weight of service activities to the detriment of manufacturing ones, provided a fertile ground for the effective introduction, adoption and diffusion of ICTs (Antonelli *et al.*, 2007, 2008).

From a historical viewpoint, the path leading to the generation and adoption of ICT emerged out of a collective and interactive process induced by

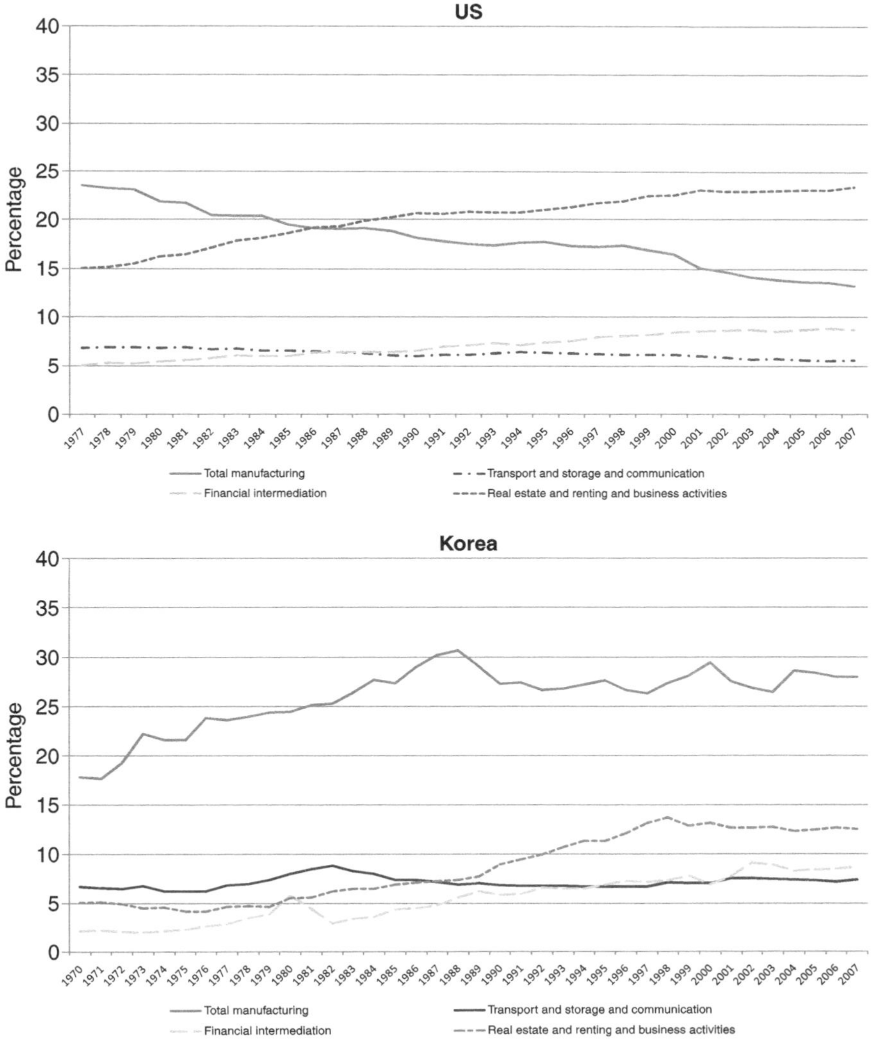

Figure 1.4 continued.

relevant changes in the economic environment. Since the late 1960s, 20 years after the Second World War, the US was experiencing a progressive erosion of its economic and technological leadership. The combined effect of the convergent catching-up of competing countries, the international diffusion of mass production and science-based technologies (Nelson and Wright, 1992) and the exhaustion of technological opportunities in chemical and engineering technologies, resulted in a strong decline of US international competitive advantage and a productivity slowdown (Griliches, 1980). This decline in performance induced myriad interdependent, sequential and creative efforts

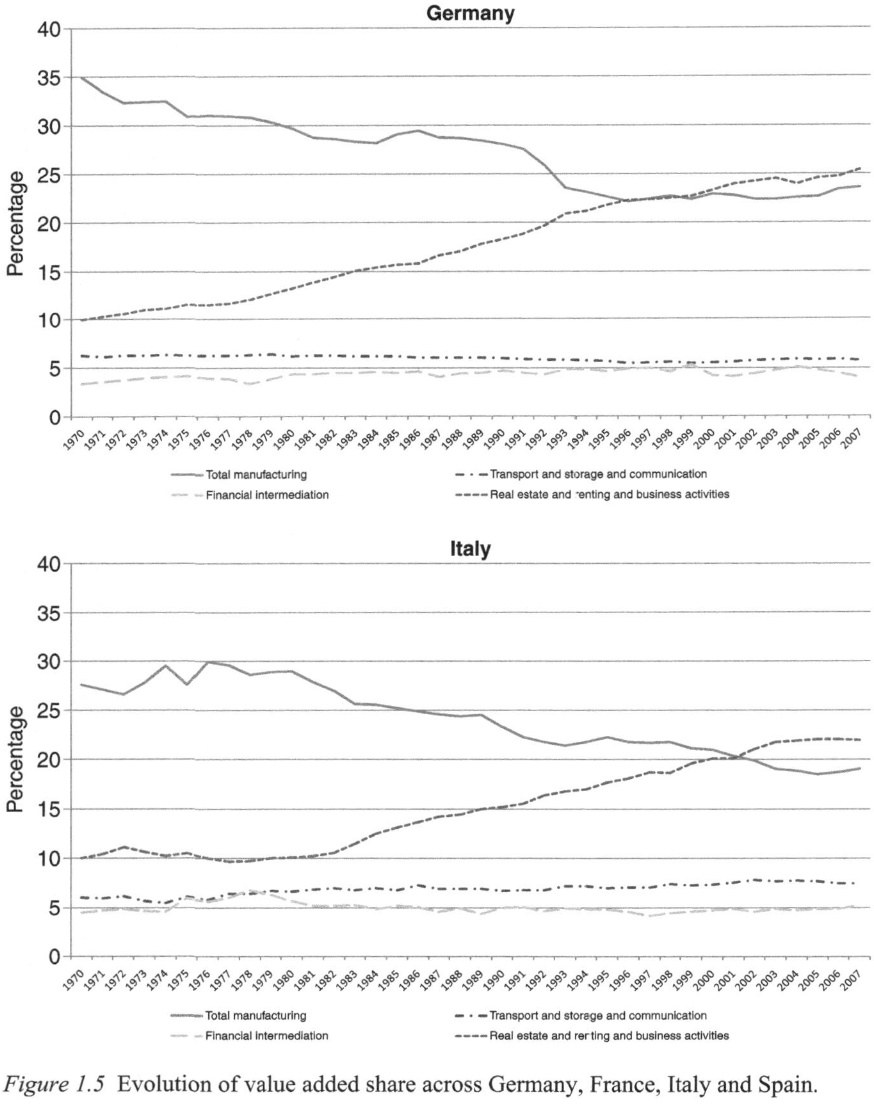

Figure 1.5 Evolution of value added share across Germany, France, Italy and Spain.

directed towards the introduction of complementary technological innovations. The main result of these developments has been the creation of a new technological system with a strong skill bias. In the decades following their introduction, ICTs have considerably improved and have slowly acquired the features of a general purpose technology (GPT). These technologies have a high degree of fungibility, that is, usable in many different contexts, strong complementarities and considerable spillover effects. Along with the improvements, the diffusion of ICT across US firms stemmed from a process of sequential, creative adoption (Lypsey *et al.*, 2005).

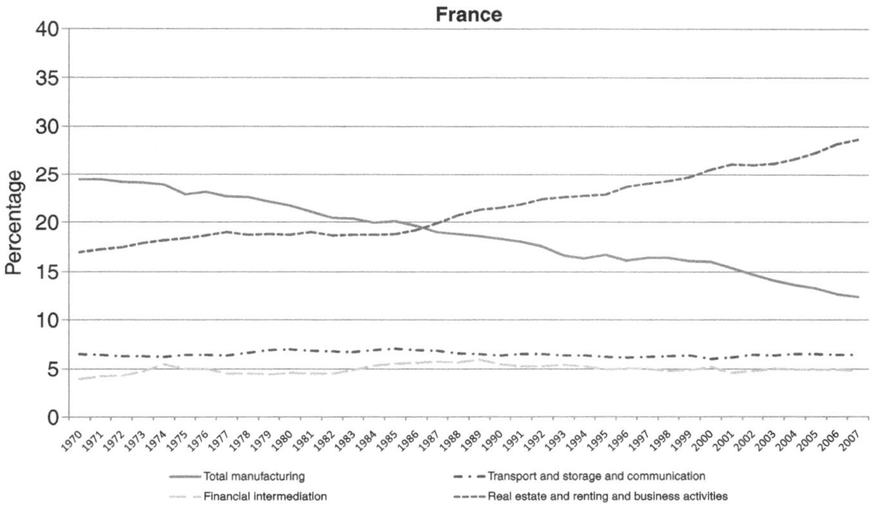

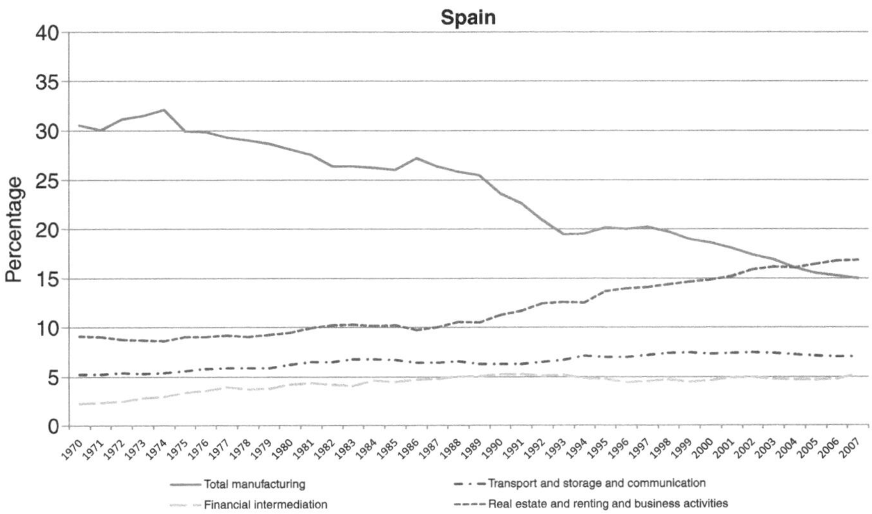

Figure 1.5 continued.

Empirical analyses of the recent unexpected US productivity surge clearly show that the main thing responsible for such growth revival in the late 1990s is technological change, in particular the introduction of new ICTs (Jorgenson, 2001). Along these lines, a rather extended body of literature offers cross-country comparisons of ICT contributions to productivity growth. While the evidence related to the diffusion of the technology is somewhat mixed, the data about productivity suggest the existence of a new process of divergence between the US and some other advanced countries in Europe (Daveri, 2002; Timmer and van Ark, 2005).

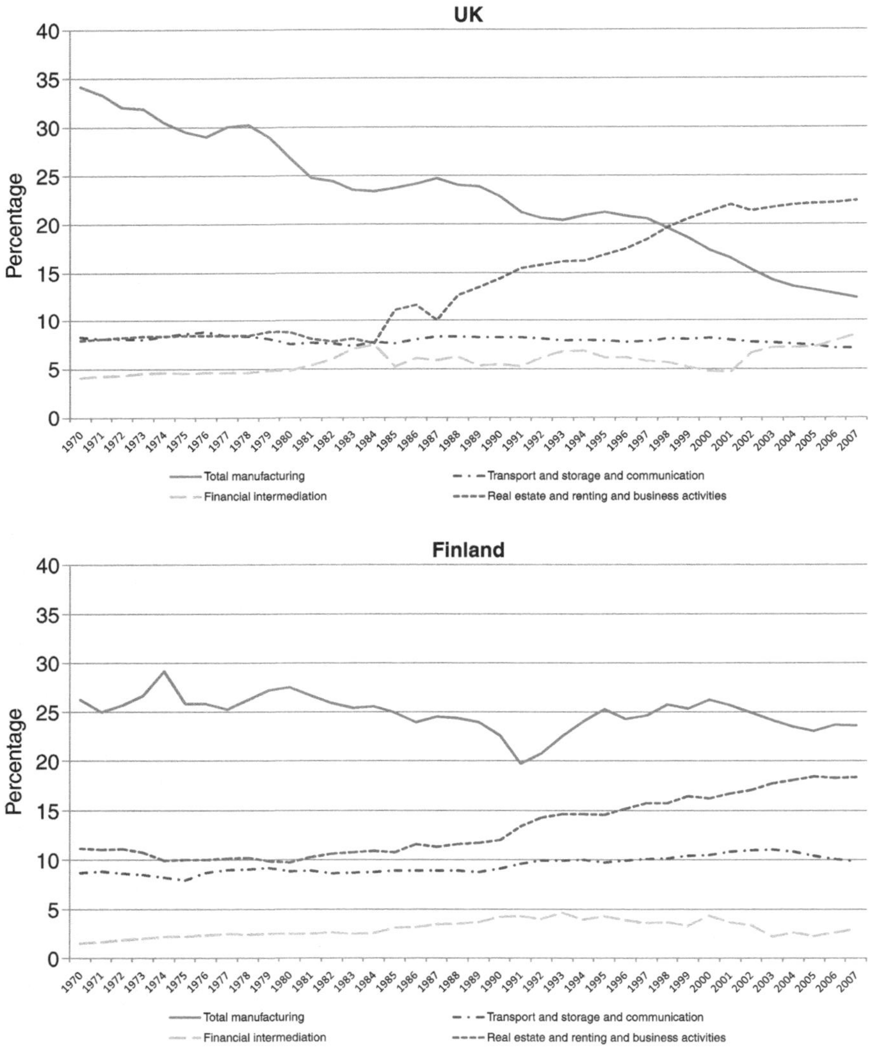

Figure 1.6 Evolution of value added share across UK, Sweden, Finland and Denmark.

The development of ICTs is clearly the result of a complex set of technological, historical, economic and institutional factors. The coupling of the evidence concerning the role of ICTs in economic development with that concerning the movement towards a service-based economy suggests that out of so many enabling elements, the process of structural change observed in the last decades provided the US economy with a competitive advantage, which translated into an increased diffusion of ICTs fostered by the rise of service activities, which in turn boosted the rate of growth of productivity. ICTs are indeed technologies showing a strong bias towards the employment of highly qualified human

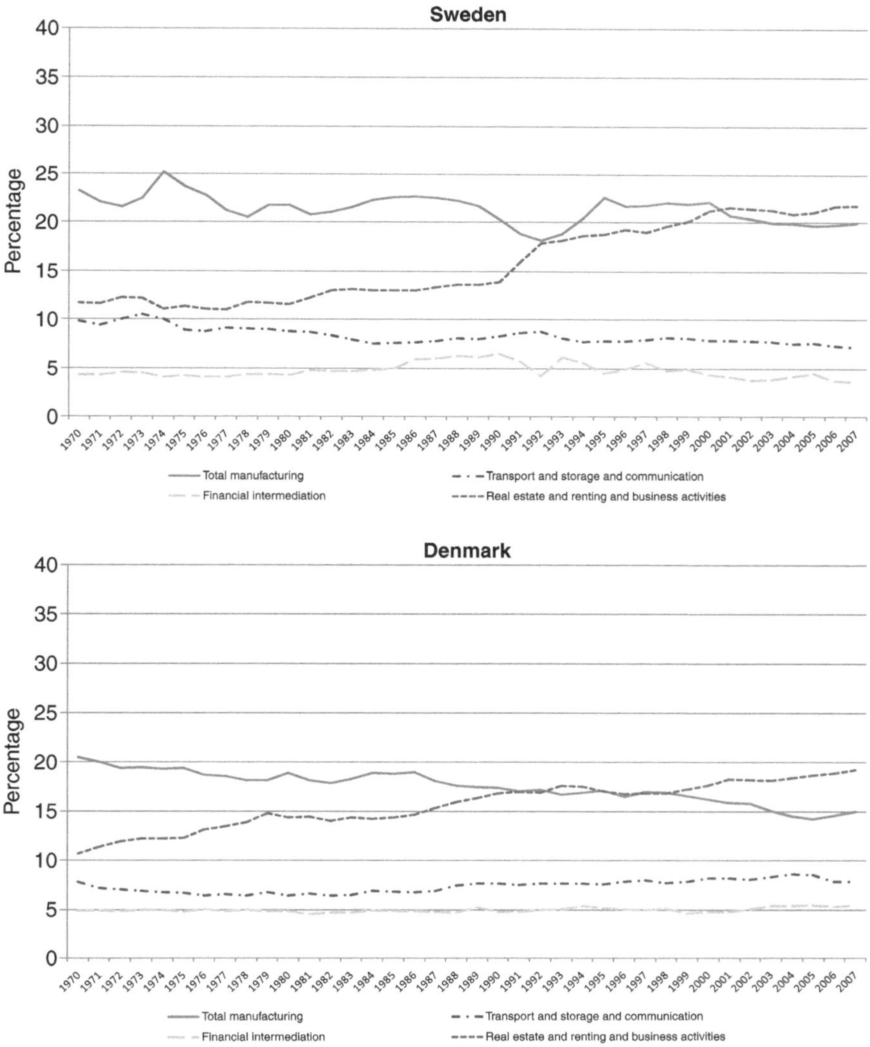

Figure 1.6 continued.

capital, able to cope with the increasing specialization in the supply of knowledge-intensive business services.

Structural change therefore interacts with institutional and economic change so as to shape the patterns of technological change. If one looks at the data on the contribution of ICTs to value-added growth across different countries[2] (Figure 1.7), it is evident that in the US it started increasing in the early 1990s and then it experienced a sudden acceleration up to the end of the decade. In the

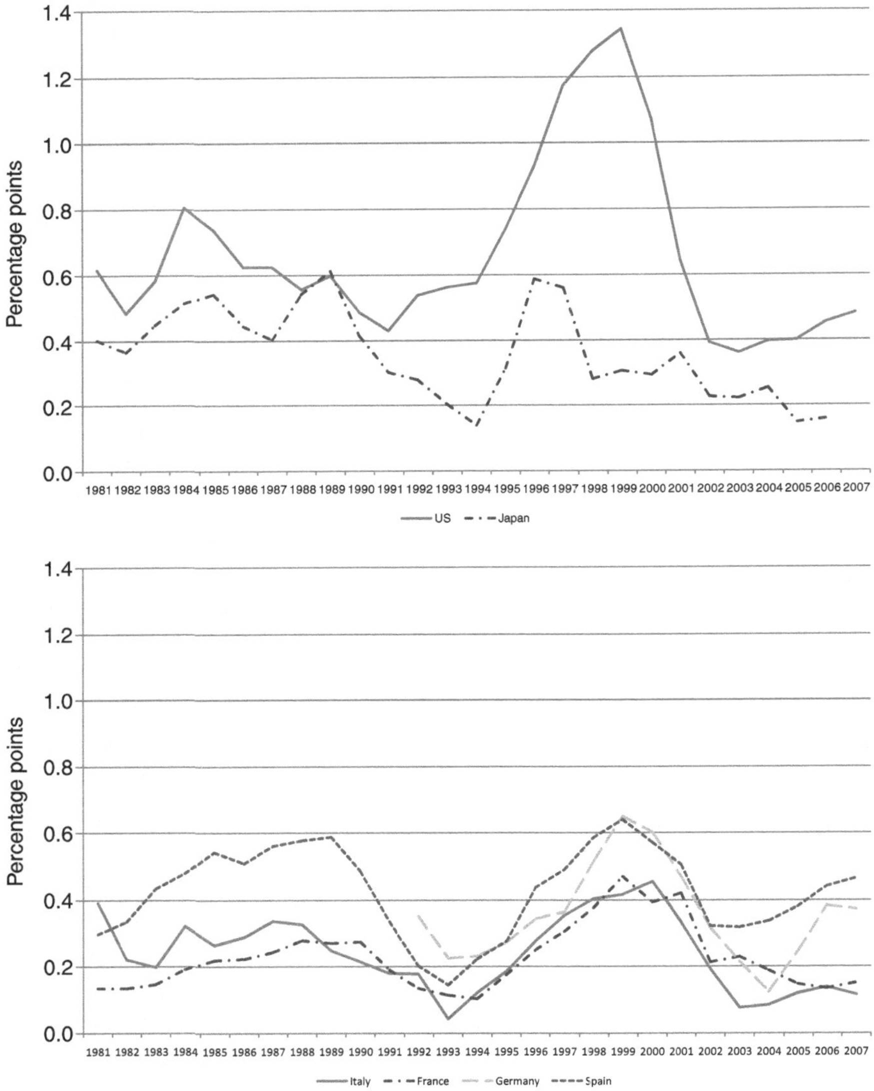

Figure 1.7 Contribution of ICTs to value added growth.

Figure 1.7 continued.

following years such contribution decreased and remained pretty stable around 0.40. This is partly due to the bursting of the NASDAQ bubble in 2000, but also to the changing pattern of contribution of ICTs to productivity growth (Jorgenson *et al.*, 2006). The data reported in Figure 1.7 refer to the contribution of ICT capital to the growth of value added, which was influenced by the fast rates of technological progress in the field, high competition and declining prices. Such dynamics characterized the phases of fastest diffusion of ICTs, but then tended to stabilize. In the most recent years it is the efficiency gains in the production of ICT-related capital that have generated the most relevant positive effects. These latter, however, are mostly reflected in productivity statistics than in value-added growth.

The data concerning Japan do not display any particular peak in the contribution of ICTs to value-added growth. We can observe a rather regular cyclical behaviour and identify a slightly decreasing trend since the second half of the 1990s. As already noted, Japan clearly lags behind the US in terms of changing industrial specialization in favour of service activities. These are in turn the main users of ICTs and are therefore the main thing responsible for their diffusion. The scarce contribution of ICTs can hence be related to the relatively low development of user services. Moreover, the global division of labour in the production of ICTs is such that only the mature modules of production process have been moved to Eastern countries in recent years, while the most promising in terms of expected returns have been retained in the US (Fransman, 2007).

In the mid diagram of Figure 1.7 we can observe the situation concerning continental European countries. First of all, it must be noted that in such areas the contribution of ICTs is far lower than that observed in the US over most of the observed period. Only in the 2000s do they appear to converge, partly as an effect of financial markets shocks. The situation is very different in northern European countries. Although we can see a pretty marked cyclical behaviour much in line with the evidence analysed so far, in the second half of the 1990s both the UK and (even more so) Sweden experienced levels of ICT contribution to value-added growth comparable to those of the US. This is very coherent with the general evidence concerning the relatively stronger weight of service sectors in these countries with respect to continental Europe.

In order to gain a more comprehensive understanding of the comparative dynamics of ICT diffusion one can look at the share of GDP expenditure for some main ICT-related goods and services. In Table 1.1 we report the expenditure for computer and office equipment, broadband and telecommunications and informatics services. These data are derived from input–output statistics provided by the OECD, concerning European countries, and the Bureau of Economic Analysis with regard to the US evidence. These figures refer to both firms and household expenditure. Of course, such a difference in data sources, and

Table 1.1 Share of GDP invested in ICTs (%)

	1995	*1996*	*1997*	*1998*	*1999*	*2000*	*2001*	*2002*	*2003*	*2004*
Computers and office equipment[a]										
Germany	3.422		3.512	3.775	3.686	3.836	3.869			
France	1.966		2.014		2.018	2.041	2.056			
UK	2.465	2.426	2.502	2.442	2.319	2.312	2.162	2.128	2.128	
Italy	1.946	1.884	1.953	2.011	1.938	1.929	1.872			
US			3.711	3.833	3.763	3.789	3.337	2.786	2.657	2.859
Broadband and telecommunications[b]										
Germany	1.681		1.756	1.913	2.272	2.337	2.401			
France	1.827		2.033		2.299	2.490	2.678			
US	2.792	3.095	3.359	3.732	3.684	3.806	3.775	4.137	4.248	
Italy	1.445	1.560	1.675	1.848	1.947	1.864	2.013			
US			1.892	2.434	2.685	2.878	3.092	3.129	3.178	3.284
Informatics services[c]										
Germany	0.956		1.177	1.422	1.493	1.603	1.912			
France	1.514		1.680		2.172	2.280	2.478			
UK	1.523	1.767	2.244	2.711	3.103	3.261	3.607	3.802	4.383	
Italy	1.208	1.359	1.403	1.541	1.738	1.717	1.893			
US			0.914	1.308	2.017	1.604	1.588	1.667	1.663	1.719

Source: Elaborations on Input–Output Data BEA and OECD.

Notes

a Product code: USA (334 + 335) and OECD (30 + 31).

b Product code: USA 513 and OECD 64.

c Product code: USA (514 + 5415) and OECD 72.

hence in product classification, makes it difficult to compare the US and European evidence. However, we can well compare the differential dynamics.

On the whole, all of the three identified products show a positive trend in the second half of the 1990s. It is also quite interesting to note that both the UK and US show a decrease in expenditure for computers in the early 2000s, while they experience an increase in the GDP share invested in broadband and telecommunications as well as informatics services. Within Europe, the UK is the country with the highest share of GDP expenditure in each product category, which is fairly in line with the fact that it appears also as the European country that mostly resembles the US dynamics of structural change. Encouraging figures are related to the French and German situation, whose path towards the service economy appears well established. Within this framework, Italy appears to have a relatively high delay in the exploitation of the potentials linked to the diffusion of ICTs, which is preventing the establishment of sound growth paths.

The evidence discussed so far emphasizes the intertwining of the processes of structural change and technological change. In particular, it provides further support for the idea that the increasing specialization in service activities has created a fertile ground for the adoption and diffusion of ICTs. Of course, this is only part of the story. ICTs are the outcome of technological efforts carried out mostly in the US and are engendered by failure-inducement dynamics set up by the oil crisis in the 1970s and the subsequent productivity slowdown. Institutional factors related to the definition of standards and the rise of venture capital also played an important role. The rise of service sectors has been crucial in setting in motion positive dynamics on the demand side, ensuring fast diffusion for such technologies and creating the condition for further development and applications (Quatraro, 2011). This synergy has paved the way for much deeper changes in advanced economies by creating the conditions for the emergence of systems based on the creation and exploitation of technological knowledge.

1.5 The emergence of the knowledge-based economy

The effects of the introduction of ICTs have been powerful. The US economy has been enjoying a new surge in productivity since 1995. The ICT industry has played a key role in this as a result of rapid technological developments in the semiconductor industry. The persistent and steep decline in the price of semiconductors has been transmitted downwards in the value chain, affecting the semiconductor-user sectors, and especially the producers of telecommunications equipment and software (Jorgenson, 2001). The productivity gains stemming from the spread of ICTs are due both to increases in efficiency in upstream industries and to the flows of creative adoptions of ICTs in downstream sectors. These technologies have enabled knowledge spillovers to the rest of the system and, as a result of intense competition, the new upstream industries have been unable to retain the full stream of benefits stemming from the new technology. This has engendered a flow of pecuniary externalities (David, 2001).

Strong US technological leadership has encouraged a new international division of labour which reversed the situation that prevailed in the 1980s. The US quickly became the main producer and user of ICTs, while the rest of the advanced countries engaged in creative adoption involving adapting the technology to the idiosyncratic conditions of their markets and industrial structures.

Because of the strong directional skill bias of ICTs a digital divide is emerging between countries that are "properly" endowed, that is, that have the "right" amount of human capital and access to the knowledge commons. These "properly" endowed countries are able to participate in the process of cumulative technological change and creative adoption. Other countries can, at best, adopt ICTs passively and enjoy fewer chances to take advantage of the new opportunities for productivity growth. ICTs are global in character because they bring about increases in productivity and efficiency, such that their adoption is profitable across a great array of products and processes and regions. Nevertheless, asymmetric effects stemming from the strong skill bias and the different endowments of human capital must be accounted for in examining these effects (Antonelli, 2003).

Since the early 1990s the adoption of ICTs has made possible the emergence of global corporations based on distributed coordination processes, selling worldwide customized products, manufactured and assembled in a variety of regions, while retaining in their home countries the skill-intensive activities. This trend is especially evident in the new service industry and, in particular, in the new knowledge-intensive business service sector (Dunning, 1993).

Specialization in new knowledge-based services that rely heavily on the quality and variety of advanced digital communication characterizes the transition to the new knowledge economy in advanced countries. ICTs are increasingly important for a wide range of knowledge services from entertainment to health and financial services to education and logistics. The advent of digital technologies changes the context in which knowledge-based services were traditionally supplied. ICTs allow remote interaction between different actors, while in traditional services, any interaction implied physical proximity. ICTs change the way in which services are delivered and used, and the way in which services are provided to final users. ICTs are crucial also to changes in the way new knowledge services are used and contributed to by final and intermediary users (von Hippel, 2005).

The centrality of knowledge in the economy has increased so much that a new branch of economics actually emerged and consolidated in the last decades, i.e. economics of knowledge, which analyses the conditions leading to the creation of knowledge on the one hand, and its economic effects on the other (Foray, 2004). Much attention has been paid in this respect to technological knowledge and to the benefits stemming from its application to production processes. In Chapter 3 we will examine in more detail the economic theories dealing with knowledge. But now, it is important to stress the close link between the increasing rates of knowledge production and the recent dynamics of structural change which led to the increase in the share of service activities and the diffusion of ICTs.

The analyses of technological knowledge have been conducted by relying on a relatively small number of indicators. They are mostly proxy variables to measure and quantify the phenomenon. The most used indicator is undoubtedly the number of patent applications filed by economic agents, which are aggregated at different levels, say firms, regions or countries. The use of patents as a proxy for technological knowledge was put forward by Zvi Griliches in the 1970s, and further developed by his students, who developed different perspectives on the exploitation of such data for economic investigation. Although patent applications show important limits that we will discuss at length further on, they nonetheless provide a useful representation of the effectiveness of knowledge-production process across the economies, at least with regard to the manufacturing sector.

In order to complete the picture of the actual relevance of structural change and its linkages with the increasing role of knowledge in advanced economies, we report in Figure 1.8 and Figure 1.9 data about knowledge production as proxied by patent applications. The former reports the number of patent applications within the Patent Cooperation Treaty (PCT). Such data are drawn by the OCED Science and Technology Indicators (OECD, 2009). In order to account for cross-country size differences the total number of patent applications has been divided by the number of employees provided by the Groningen Growth and Development Centre so as to obtain a relative measure of patenting activity, which is more useful for the purpose of comparison. In the top diagram we compare the four main players in the world economy. One can immediately observe that the rate of creation of knowledge is positive over the whole observed period. In particular, in the US there is a significant boost in the second half of the 1990s which becomes even more evident in the 2000s. This represents interesting, though descriptive, empirical evidence of the correspondence between the gradual dominance of service over manufacturing activities, which became evident in the US already in the 1990s, the diffusion of ICTs and the increasing relevance of knowledge-production activities. The evidence on Japan and Korea is as much interesting, as we can observe a sudden acceleration for the former since the second half of the 1990s while in the latter this occurs in correspondence with the end of the same decade. The EU-15 aggregate shows a smoother dynamics in which the growth rate, differently from the other countries, decreases in the early 2000s.

In order to better understand the European evidence, the mid diagram of the figure shows the evidence of continental European countries. Even controlling for cross-country differences, patenting activity in continental Europe seems to be far lower than in the US or Japan. As expected, the countries showing the best performances are Germany and France, which experience a marked acceleration in the second half of the 1990s. Indeed these appear to be also at the forefront in the changing specialization in favour of service activities, despite the relatively high share of manufacturing that still persists in Germany. The Italian and Spanish evidence confirms instead the relative delay of these two countries. The bottom diagram shows finally the evidence concerning north European countries.

These figures are comparable to the US and Japanese evidence also in terms of magnitude. This is especially true with regard to Denmark, Finland and Sweden, while the UK shows figures significantly below those of such countries. The best performing country here is Sweden, closely followed, and actually overtaken in 2001, by Finland. The acceleration in the rate of creation of new knowledge in these two countries occurred in the first half of the 1990s, while it can be observed towards the end of the decade in the case of Denmark.

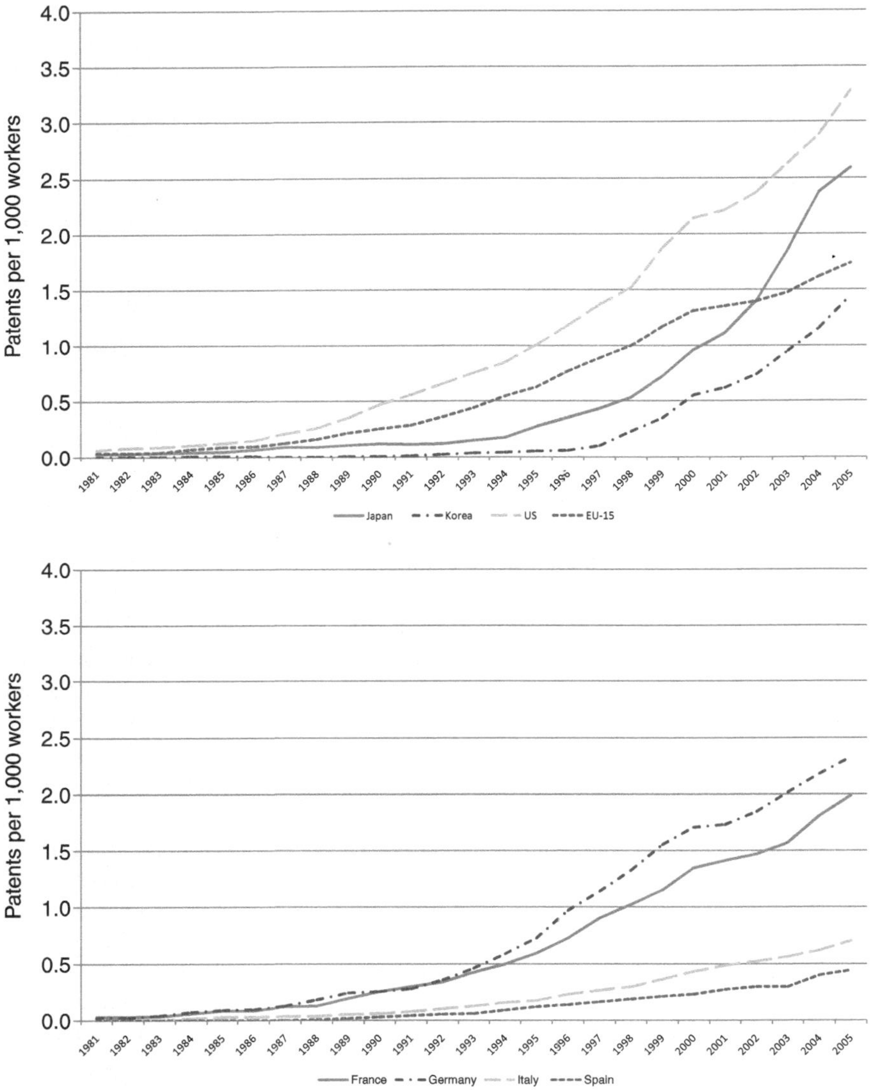

Figure 1.8 Dynamics of patents per 1,000 employees.

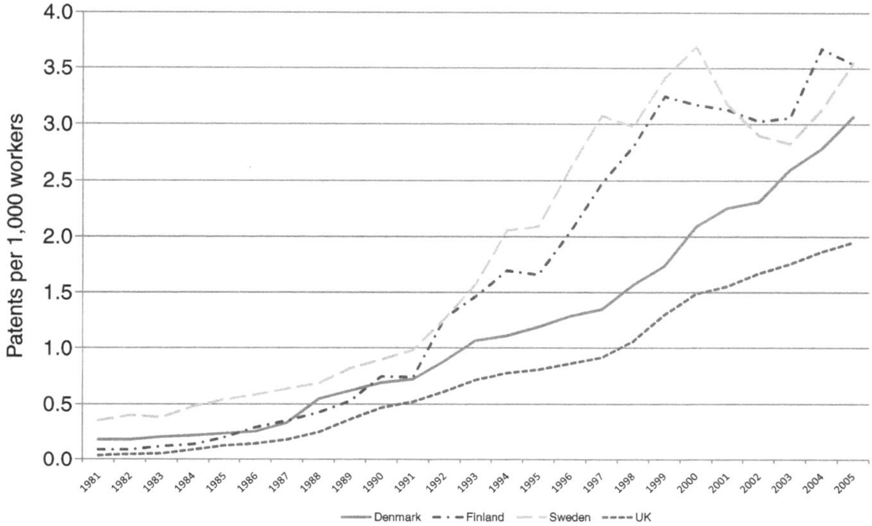

Figure 1.8 continued.

The data on knowledge creation provide therefore a useful complement to the understanding of the implications of the recent dynamics of structural change which took the shape of a transition from manufacturing to service-based economic systems. Further interesting information can be obtained by looking at the share of patent applications within the PCT which are related to ICTs. In Figure 1.9 we report the dynamics concerning the US, Japan, Korea and the EU-15. We can observe that Japan and the US retain the highest relative levels of ICT-related patents for almost the whole observed period. In particular, the US became dominant in the second half of the 1990s, i.e. in the period in which its productivity dynamics was mostly driven by the creation, diffusion and exploitation of ICTs. Such a position was retained until 2002, when it was overtaken by Korea and Japan. In the mid diagram the data on continental European countries suggest that again France and Germany have a significant advantage with respect to Italy and Spain. Germany outperformed France only in the second half of the 1990s, while in the rest of the observed period the curve concerning France was constantly above those of the other countries. It must be noted, however, that the share of ICT-related patents never exceeded 35 per cent in central European countries, while in the case of the US and Japan the peak was reached at 45 per cent. The bottom diagram provides instead the evidence about north European countries, showing that since the early 1990s Finland has taken the lead in this group, growing constantly until 2007 at a rate such that in 2007 about 60 per cent of its patents were in ICT-related technologies. Sweden follows Finland, although with significantly lower values, while the UK and Denmark are characterized by dynamics which are not particularly relevant.

These data allow us to gain a more comprehensive picture of the structural transformation ongoing in advanced capitalist economies, so as to include the increasing relevance of knowledge production and, in particular in the second half of the 1990s, of technological knowledge related to the development and creative adoption of ICTs. We are now in the position to draw some preliminary conclusions that will serve as a basis to develop the heuristic framework in the next three chapters.

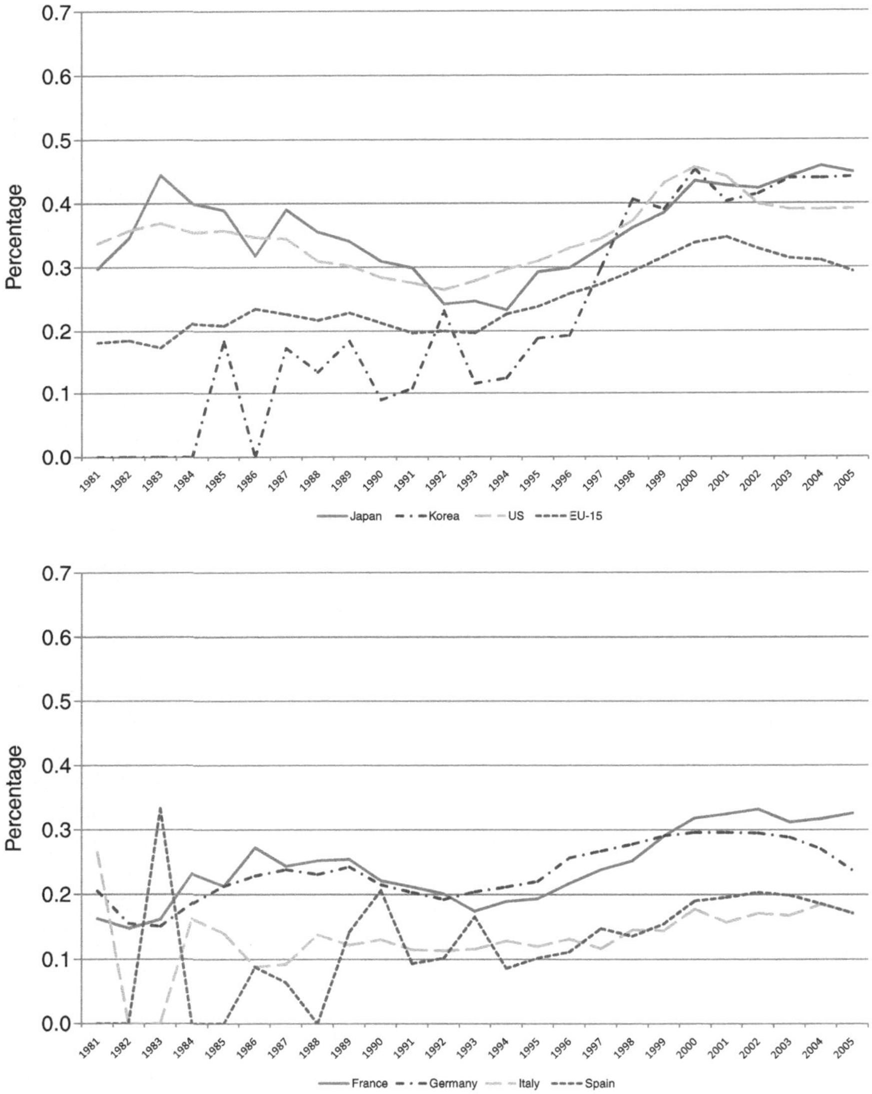

Figure 1.9 Share of ICTs patents on total applications.

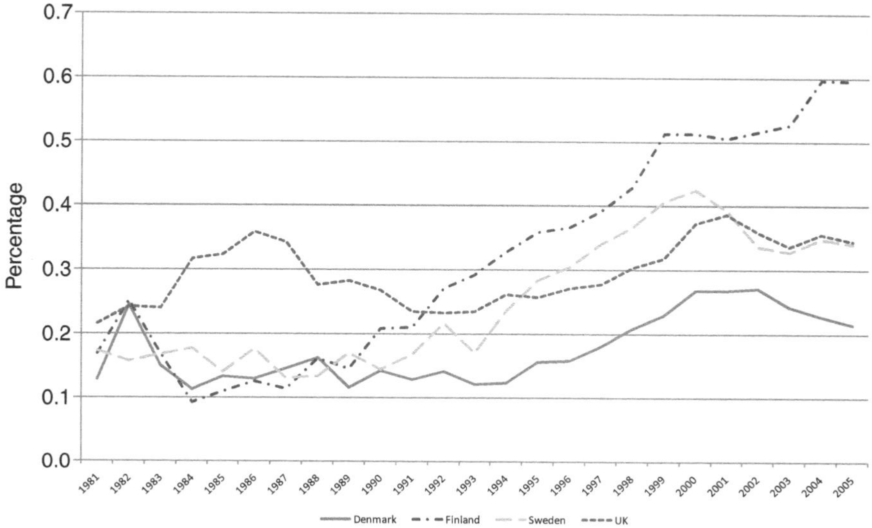

Figure 1.9 continued.

1.6 Conclusions

The analysis of industrial and technological dynamics allows us to outline an interesting picture representing the evolution of structural features of most advanced economies. Three main groups of countries clearly emerge, on the basis of the relative advancement in the transition process towards a service- and knowledge-based economy. The US appears to be the leader in this context, closely followed by Japan and, more recently, Korea. North European countries appear to be characterized by a somewhat long-lasting tradition of high service share in the economy, though the data on knowledge production suggest the establishment of a significant active role in this field only in the late 1990s. The central European countries show instead a quite worrying delay, which can partly explain why the persistence of US leadership is being challenged nowadays by some Far Eastern countries, but not at all by European ones. Of these latter, France and Germany are clearly far better positioned than Italy and Spain, in which industrial activity seems to be still too much dependent on the evolution of manufacturing activities. The low levels of knowledge production suggest moreover that such activities are hardly specialized in the production of high-tech goods, and are more likely to involve mature and labour-intensive sectors which are now increasingly developing in areas characterized by lower wage rates for unskilled labour.

The evidence discussed in this chapter suggests not only that structural change deserves to be properly investigated in order to fully understand differential dynamics of economic development. The matching of economic variables with technological indicators also calls for a more extended approach to the analysis, involving also the appreciation of the role of innovation dynamics. The integration of innovation and knowledge into the picture can be far-reaching, in that it lends itself to a broader declension of the concept of structural change. In particular, the recent developments on the dynamics of collective knowledge generation through the recombination of dispersed and fragmented knowledge provide the basis to bring into the framework of analysis the structure of networked agents as well as the structure of knowledge bases. Each structure is likely to be characterized by its own architecture, which in turn is likely to have effects on the actual performances on the working of whole system. A complex chain of feedbacks and mutually enforcing dynamics can therefore take place in this direction. The location aspects play a key role in this respect, in that they are likely to introduce powerful constraints to the way such structures evolve over time as well as to the way each structure exerts its influences on the other ones.

The next part of the book will be devoted to the development of a path moving from the traditional theories of structural change, going through the integration of the theories about innovation and knowledge production, so as to achieve a temporary synthesis based on the heuristic tools provided by the complex systems theory and its applications to the economic domain.

Notes

1 We used the EU KLEMS database, available at www.euklems.net.
2 Source: EU KLEMS database, provided by the Groningen Growth and Development Centre.

Part II

The theory

2 Structural change and the long-run dynamics of economic growth

2.1 Introduction

The dynamics of structural change do not represent a phenomenon limited in time and space, but rather look like a continuous process ongoing in different geographical and industrial contexts, linked to the phases and cycles of economic development. The empirical evidence provided in the previous chapter speaks for this, and shows how uneven are growth rates across sectors and, for the same sectors, across different countries.

Since its origins, the concept of structural change has related the changes in the sector composition of the economy. It has then been enriched by using it to denote additional phenomena like changes in firms' size distribution or institutions. While there is a wide body of empirical evidence discussing this issue, analytical models are relatively less numerous. Some models are grounded on the supply side, focusing on the asymmetric dynamics of labour productivity, while other models draw upon a demand-side approach, mainly based on the inclusion of nonhomotetic preferences in neoclassical growth models or on Engel's law. Both types of approaches share the same limitations, in that technological change is recognized as an important aspect of the process, but it is evoked as exogenous. Such drawbacks have been addressed by evolutionary models based on replicator dynamics, which have the clear merit of drawing the attention of the intrinsic relationships between the analysis of structural change à la Kuznets and the Schumpeterian analysis of technological change.

This chapter elaborates the path driving from the antecedent scholars dealing with structural change to the articulation of the intertwining with the study of technological change. Section 2.2 explores the origins of the analysis of structural change in economics, moving from Smith through Marshall and Young. Section 2.3 illustrates the so-called three-sector hypothesis, emphasizing the consequences in terms of convergence across regions and countries. Section 2.4 provides an overview on the analytical approaches to the analysis of structural change, while Section 2.5 articulates the link between structural change and technological change, by showing similarities and complementarities between Kuznets and Schumpeter. Finally Section 2.6 provides provisional conclusions.

2.2 The origins of the analysis of structural change in economics

The identification of the origins of the analysis of structural change in economics is not a very easy task, for at least two sets of reasons. First, it is difficult to detect a unique meaning of structure and structural change in economics. Second, the utilization of the term "structural change" is relatively recent, and it is likely that key authors in economic science have dealt with structural change without having explicitly mentioned it.

We have already noted in the first chapter that the term structural change can indicate different research contexts in economics. A former systematic analysis of the different meanings that this expression can take in economics can be found in Machlup (1963). He indeed provided an extensive list of the uses of terms, assessing also the degree of clarity of the utilization. The most common use of the term concerns the different arrangements of productive activity in the economy, with particular reference to the different distribution of productive factors among various sectors of the economy, occupations, geographic regions, types of product, and so on and so forth (Machlup, 1963). The key reference in this respect is the Nobel Laureate Simon Kuznets, who dedicated most of his research activity, which will be the object of detailed analysis in what follows, to the analysis of the changing distribution of employment across industries and the relationship between stage of development and the industrial composition of national economies (Kuznets, 1930, 1973).

However, some more remote contributions dealing with similar issues can be found, put forward by the founding fathers of economics. For example, Adam Smith's *Wealth of Nations* (1776) articulated as a main hypothesis that the increase in the final demand for goods engenders the division of labour, by creating new branches of activity. Moreover, Smith explicitly stated that *those countries which have successfully developed a specialization in manufacturing activities are those that most reap the benefits stemming from the division of labour*, i.e. efficiency gains: "The most opulent nations, indeed, generally excel all their neighbours in agriculture as well as in manufactures; but they are commonly more distinguished by their superiority in the latter than in the former" (Smith, 1776, p. 111). In Smith the industrialization process is strictly linked to the division of labour, which in turn contributes to the accumulation of new skills and competences. The division of labour is also at the basis of technological change, channelled both by learning dynamics (the increase of dexterity) and the creation of new machineries, often stemming from "the ingenuity of the makers of the machines, when to make them became the business of a peculiar trade; and some by that of those who are called philosophers, or men of speculation" (Smith, 1776, p. 9). Thus, in the *Wealth of Nations*, poor countries are those mainly specialized in agriculture activities, whereby the division of labour is limited by the nature of the tasks to be carried out and therefore can yield very limited productivity gains. On the contrary rich countries, what we would call today "developed countries", are those specialized in manufacturing activities, better suited to be articulated in different tasks so as to increase production efficiency. It must be also noted that in Smith structural change does not concern just the sectoral composition of economic systems, but also the organization of production within organizations.

As Silva and Teixeira (2008) note in their bibliometric survey on structural change, besides Smith, there are also other "classical" economists who dealt with the changing composition of economic structure, although not explicitly. Ricardo (1817) noted how sustainable output growth requires the growth of production factors. As the classical production factor, i.e. land, is limited, sustainable output growth can be attained only by substituting produced for non-produced inputs, and as a consequence, by managing the shift from the specialization in agriculture to manufacturing activities. On more analytical grounds, Quesnay (1758) and Marx (1885) also provided contributions to the understanding of the change of economic structure. The former explored the interdependencies among industrial sectors, proposing a description of the analytical structure of the economy based on the concept of "natural proportions" between sectors. In a similar perspective, Marx distinguished between constant and variable capital and argued that the increase in the ratio between the former and the latter implies a re-proportioning of the various commodities produced.

Classical economists provided therefore an early treatment of the intrinsic change of economic structure typical of capitalist economies, and identified a clear pattern directed towards the increasing weight of manufacturing activities with respect to the agriculture ones. Surprisingly enough, the most recent surveys on structural change (Kruger, 2008; Silva and Teixeira, 2008) neglect a fairly important contribution in this sense, coming from another key author for the discipline: Alfred Marshall. In particular, *Industry and Trade* (1919) anticipated most of the arguments that would have been put forward by Kuznets (1930) and Joseph Schumpeter (1939). The core of Marshall's argument is that trade patterns deserve to be investigated in that they reflect a country's industrial leadership. The context to which the analysis applies is of course one of international division of labour, in which the reduction of transport costs play a key role in allowing for the extension of final markets to foreign countries. However, the author stresses that his line of reasoning also applies to the analysis of trade flows between regions or even smaller areas, for which statistics are in any case hardly available. In this perspective, the balance of payments represents a useful source of information, which of course is not exhaustive, but which helps to identify where to concentrate the interest of the researcher.

According to Marshall, advances in industries in which the country already possesses a competitive advantage are likely to strengthen international trade. If a country already shows an excess output in an industry which is absorbed by international markets, advances in that industry will improve the production process, making the final goods even more attractive for foreign buyers. On the other hand, advances in an industry in which the country is outperformed by other countries should reduce the trade flow, as the improvements will lead the country to reduce imports of the good produced.

The pattern of industrial specialization is therefore a key aspect influencing the trade between nations. However, competitive advantages are not supposed to characterize the same industries forever, and accordingly industrial leadership of countries is likely to follow the evolution of the main industries they are specialized in. This does not imply necessarily the switch to different activities, which can be eventually attained only as a result of a very slow process. An alternative to cope with the challenges coming from emergent countries which are likely to follow a delayed development path similar to those of the advanced ones is the introduction of improvements, not only technical, to increase the efficiency of production processes. In this direction, investment in the education sector turns out to be crucial for countries specialized in activities which are already ahead in the stage of development. Marshall provides an account of these dynamics hardly relying on strong statistical bases, but providing an interesting description of the patterns of evolution of industrial leadership and of trade flows in the UK, France, Germany and the US. In this sense his work can be considered a dense contribution in business history, in which changes in economic structure affect different dimensions, ranging from foreign trade to the organization of production.

Marshall's emphasis on division of labour also marks an important difference from Adam Smith's articulation of the concept. As Marshall makes clear in *Principle of Economics* (1890), Smith's argument is mostly focused on the dynamics and the effects of division of labour within firms' boundaries. Thus, the benefits stemming from the division of labour, channelled by the increased dexterity, saving the time that should be devoted to passing from one task to another, and the introduction of new machineries, these are all related to the internal economies of the firm.

Marshall pushes the argument further and articulates the analysis of the division of labour at the system level, anticipating the analysis of the sources of industrial differentiation in local contexts. The key point in this respect lies in the analogy that Marshall put forward between physical and social organisms, of which industries are clear exempla: "the development of the organism … involves an increasing subdivision of functions between its separate parts on the one hand, and on the other a more intimate connection between them" (Marshall, 1920 [1890], pp. 200–201). Such a differentiation implemented through the division of labour at the industrial level is also likely be characterized by a higher degree of integration, i.e. interconnections between the different parts of the industrial organism: "Each part gets to be less and less self-sufficient, to depend for its wellbeing more and more on other parts, so that any disorder in any part of a highly-developed organism will affect other parts also" (Marshall, 1920 [1890], p. 201).

Marshall's analysis of external economies is grounded on localized industries, i.e. in contexts characterized by the concentration of business activities of similar character. In these contexts one can observe the dynamics of differentiation and integration at work. Although the reasons behind the specialization of specific areas or regions in some industrial activity may be diverse, the localization of industry is likely to engender system dynamics which benefit firms operating therein. Such dynamics consist of Marshall's well known externalities. An important part of these consists of the fact that "subsidiary trades grow up in the neighbourhood, supplying it with implements and materials, organizing its traffic, and in many ways conducting to the economy of its material" (Marshall, 1920 [1890], p. 225). The localization of industry therefore enhances the division of labour at the industry level as a response to increasing volume of output stemming from increasing demand. Horizontal and vertical diversification coexists, the latter taking advantages also of the opportunity for upstream firms specialized in a small part of the production process to supply many downstream firms operating in different and yet technically similar industries, so as to make the most efficient use of the highly specialized machinery. The extent of the market therefore has effects on individual firms which are different from those showing up at the industrial level. The increase of the volume of production always increases the level of external economies, increasing the general efficiency of the system.

These arguments have been further developed by Allyn Young (1928), who grafted Smith's analysis of division of labour onto a dynamic Marshallian framework in which specialization leads to speciation of new industries closely intertwined with one another. According to his analysis, increasing returns are likely to generate economic advantages in the context of roundabout methods of production. Such advantages are largely similar to those arising from the division of the labour, but Young argues that "we look too much at the individual firm or even ... at the individual firm" (Young, 1928, p. 531). In order to grasp the effects of the economies generated by increasing returns one needs to shift attention from large-scale to large production, by considering the overall output of the economic system rather than the dimensions of the market with which the single firm is confronted. In this direction "increasing returns are reflected in changes in the *organisation of industrial activities*" (Young, 1928, p. 537; emphasis added).

In line with Marshall, Young stresses that the main effect of the growth of production is industrial differentiation, which translates into the diversification of the production of both final goods and intermediate goods. This latter phenomenon is particularly relevant in modern economies, in which "over a large part of the field of industry an increasingly intricate nexus of specializing undertakings has inserted itself between the producer of raw materials and the consumer of the final product" (Young, 1928, p. 538). This process, when originated by the increase of the volume of production, generates increasing returns: "In so far as it is an adjustment to a new situation created by the growth of the market for the final products of industry the division of labour among industries is a vehicle of increasing returns" (Young, 1928, p. 538).

Although Young's emphasis on the relationship between the dimension of the market and industrial differentiation may be interpreted as a dynamic driven by consumer behaviour, it must be noted that on the contrary the advantages of increasing returns become manifest as long as roundabout methods of production are at stake. This is explicitly underlined by the author: "the largest advantage secured by the division of labour among industries is the fuller realising of the economies of capitalistic or roundabout methods of production" (Young, 1928, p. 539). In other words, high growth rates of an industry are likely to make more effective the articulation of the production process in different tasks carried out by separate firms. The disentangling of the phases of production realized in such a way may give rise to new industries which are obviously complementary (or auxiliary) to the original one. An increase in the demand for the final good produced by the original industry has positive effects, economies of second order, which translate into the increase in the derived demand for the products supplied by firms in auxiliary industries, which can in this way fully exploit the capacity of their production process. This in turn allows the lowering of unit costs of production, which translate into lower prices for downstream firms.

Young therefore firmly believes in the necessity to look at industrial operations as an interrelated whole, in the same way that Marshall represented industrial activity as an organism made of separate yet complementary functions. Regarding the understanding of the process of structural change, he clearly has the merit to have extended Marshall's analysis of external economies so as to investigate the benefits stemming from increasing returns generated by the division of labour at the industrial rather than the firm level. The creation of new industries, or new branches of economic activities, is thus fully endogenized, even in absence of technological progress, altering the sectoral composition of the economic system. However, since Smith, through Marshall and Young, economic life is ruled by market transactions. Such an assumption becomes more and more difficult to hold as industry moves towards organizations characterized by increasing vertical division of labour, whereby industrial differentiation is dominated by the emergence of auxiliary sub-industries showing a high degree of integration. A clear problem of coordination arises, which is not taken up by the authors. As Richardson (1972) suggests, the higher the complementarity between economic activities, the more difficult it is to rely on market transactions as a coordination device. Cooperation is more likely to manage successfully the ex-ante matching of production plans of firms operating in complementary sectors. This represents another important feature of structural change, which is strictly related to the evolutionary patterns of industrial development.

2.3 The analysis of structural change in the 1930s: the three-sector hypothesis

While the contribution of "classical" economists provides a former and implicit treatment of the dynamics of structural change, it was only in the 1930s that major economic contributions became explicitly focused on the analysis of the process of industrial evolution and the link between economic growth and industrial leadership. The development of the main industries characterizing the sectoral specialization of countries became a key field of inquiry to understanding the changing distribution of the economic leadership.

The approach to the analysis of structural change developed in this period is known as the three-sector hypothesis. The empirical accounts used to partition the economic system into three main aggregates: the primary sector, roughly corresponding to agriculture, fishery and forestry; the secondary sector, which produces consumption and investment goods by combining capital, labour and intermediate goods; and the tertiary sector, providing business services. This line of inquiry postulates a systematic succession of the development of the three main sectors of the private economy.

Key authors in this framework are Arthur Burns (1934), Allan Fisher (1939) and Simon Kuznets (1930). The last of these is clearly the one having proposed a detailed analysis of such dynamics, and he can surely be identified as the founder of the strand of empirical analysis of structural change. The building block of Kuznets' approach is the growth retardation hypothesis, which was articulated in his 1929 article in the *Journal of Economic and Business History* and fully developed in his famous book, *Secular Movements in Production and Prices* (1930).

The theory of growth retardation states that industry growth rates are declining over time, and then that industries whose period of development comes later are likely to overtake the mature ones. This implies that one would observe an alternation of leading industries, and of leading countries as well. Such diversity across industries generates a process of change in the economic structure of production, in terms of relative composition of activities. Differential growth rates across branches of an industry are hence likely to create structural change.

The core of the growth retardation theory can be grasped by reporting the following passages, presenting the two basic points. First of all:

> if we single out the various nations or the separate branches of an industry, the picture becomes *less uniform*. Some nations seem to have led the world at one time, others at another. Some industries were developing rapidly at the beginning of the century, others at the end. Within single countries or within single branches of industries ... there has not been uniform, un-retarded growth.
>
> (Kuznets, 1930, p. 3; emphasis added)

The *unevenness of growth rates* hence appears to be the first pillar of the theory. The intertwining of cross-industry and cross-country dimensions is of particular relevance. That is to say, the performance of a country is strictly related to industry dominating within that country, and to its relative stage of development.

Kuznets acknowledged these dynamics. Indeed one of the six characteristics of modern economic growth he proposed was the high rate of structural transformation of the economy. He wrote:

> Major aspects of structural change include the shift away from agriculture to nonagriculture pursuits and, recently, away from industry to services; a

> change in the scale of production units, and a related shift from personal enterprise to impersonal organization of economic firms.
>
> (Kuznets, 1973, p. 248)

Hence the process involved not only the distribution of employment across sectors, but also the dimensional distribution of productive units as well as their organizational forms. Elsewhere the author argued that the shift in the structure of production, and the stream of technological innovation, have been at the core of the economic history of the US (Kuznets, 1977).[1]

Besides unevenness, we find the *reduction of industry growth rates over time*. The factors underlying the dynamics of industrial growth can be grouped into three classes: (1) population growth, (2) changes in demand and (3) technical progress. First, population growth and economic development are mutually influencing. It is just another productive factor. The tendency towards decline in the rate of population increase in advanced countries would hence support the evidence of declining industrial growth rates. Second, consumer demand represents a retarding force, since there are definite limits to the amount of a commodity a person can consume. Industrial growth is thus retarded by the saturation of the total volume of consumers' demand.

However, Kuznets put the main emphasis on the role of technical progress in explaining the slackening of industry growth rates. He devoted a large part of the first chapter of *Secular Movements* to articulate the dynamics supporting the view of a slackening rate of technological change over time. He embraced the view expressed by Julius Wolff, whereby

> every technical improvement, by lowering costs and by perfecting the utilization of raw materials and of power, bars the way to further progress. There is less left to improve, and this narrowing of possibilities results in a slackening or complete cessation of technical development in a number of fields.
>
> (Wolf, 1912, pp. 236–237, quoted in Kuznets, 1930, p. 11)

According to Wolff's law, technological opportunities within a given industry, or within a branch of an industry, are likely to be exhausted over time. Since technological change is the main engine of economic growth, the slackening of technical progress determines the slow-down of industry growth rates. The decelerating industry is in turn likely to exercise a retarding effect upon the faster growing industry.

The dynamics described by Kuznets by no means lead to economic paralysis. Progress, and hence economic growth, slows down unless there is a new radical breakthrough that is likely to create a new industry, having unexploited potential for development, and in turn feeding economic growth. The view expressed by Kuznets is one in which technological knowledge creates not only the conditions for inventions and innovations exploitable in the production process, but also the conditions for generation of further technological knowledge, providing an earlier account of a self-enforcing mechanism in which economic growth is an autocatalytic process (Metcalfe, 2003).

As emerges in Kuznets' examples comparing the UK, Belgium and Germany, such a situation is due to the fact that

> as we observe the various industries within a given national system, we see that *the lead in development shifts from one branch to another*. The main reason for the shift seems to be that a rapidly developing industry does not continue its vigorous growth indefinitely, but slackens its pace after a time, and is overtaken by industries whose period of development comes later.
>
> (Kuznets, 1930, p. 5; emphasis added)

As noted by Syrquin (2010), in the 1930 book almost all the ingredients of Kuznets' approach to the analysis of structural change, which would eventually be enriched in subsequent contributions, point to stressing further dimensions of the process, like cultural change, income distribution and institutions (Kuznets, 1973, 1989). However, despite the richness of the analysis developed within this research programme, Kuznets' contribution has remained somewhat neglected for quite a long time, with the only exceptions being the interesting empirical efforts by Moshe Syrquin and Hollis Chenery, aimed at expanding his approach so as to investigate the development patterns of developing countries in the post-war period (Syrquin, 1988; Chenery, 1960; Syrquin and Chenery 1989).

2.3.1 Implications: structural change and convergence

An interesting implication of the three-sector hypothesis and the growth retardation theory is that cross-country differences may well be the result of differences in the economic structure. A country wherein the leading industry is a relative young one is likely to enjoy higher growth rates than a country in which the economic activity is led by a mature industry. Metcalfe (2003) rephrased this principle in terms of economics ecology, arguing that the extent of retardation[2] is determined by the growth rate of industry output in relation to the growth rate of its niche. This is to say that the lower the level of initial output, the faster the branch will fill its niche, engendering a relative higher growth rate of output.

Linking the economic performances of a country to the development stage of its leading industry allows for relating the hypothesis of productivity convergence to the extent to which economic, social and political forces are able to stimulate and sustain a process of structural change, in which the younger and faster-growing industries are favoured.

The classical convergence hypothesis stems from Solow's model of economic growth (Solow, 1956, 1957). Because of the stability of the equilibrium, economies with different endowments of capital per worker have to reach the same steady-state growth rate. This would suggest that economies with lower endowment of capital per labour are expected to grow faster than economies with a higher endowment.

The first empirical test of the convergence hypothesis can be found in Baumol (1986), who carried out an analysis on 16 OECD data, finding evidence of productivity convergence. In the same year, Moses Abramovitz published the famous paper "Catching up, forging ahead, and falling behind" in the *Journal of Economic History*, using, like Baumol, Maddison's data to analyse economic dynamics during the quarter century following the Second World War. The main point was that countries in the "industrialized" West had been able to take advantage of unexploited technology, mainly consisting of methods of production and organization already in use in the US. Follower countries hence had the opportunity to catch up with the leader, i.e. the US. Such a convergence, according to which productivity growth rates tend to vary inversely with productivity levels, varied from period to period, and across countries.

The link with the work of Kuznets, who was Abramovitz's teacher, appears immediately when one reads:

> These views about post war following and catching up suggest a more general hypothesis that the productivity levels of countries tend to converge. And this in turn brings to mind old questions about the emergence of new leaders and the historical and theoretical puzzle that shifts in leadership and relative standing present.
>
> (Abramovitz, 1986, pp. 385–386)

The "old questions" clearly refers to unevenness of growth rates across countries and industries.

The issue is strictly related to growth retardation, insofar as it is argued that being backward in level of productivity carries a potential for rapid advance. Following Kuznets (1930), one can define a ceiling level of productivity, and measuring the growth rate as proportional to the distance of the country, or industry, from the ceiling. The growth path for productivity hence takes an S-shape, being governed by a logistic process. Abramovitz went further, by proposing that having grasped the opportunity for catching up, the group of followers went into the retardation of productivity growth that most advanced countries have suffered since 1973.

The influence of Kuznets in Abramovitz's treatment of convergence can also be found in his discussion of the concept of social capability, a term referring to the set of societal characteristics which allow potential for rapid growth to be realized. "A country's potential for rapid growth is strong if it is technologically backward but socially advanced" (Abramovitz, 1986, p. 388). The set of factors constituting social capability – i.e. level of education, organizational experience to manage large-scale production and access to capital markets – are complementary to the obstacles to the spread of the industrial system identified by Kuznets (1954). Three different obstacles were proposed: the very specific nature of scientific and technological knowledge, arising in response to problems linked to the idiosyncratic conditions of production and as such difficult to adapt elsewhere; the disturbing action of pioneer countries exerted to retain their economic superiority; and the need for realizing complementary social and institutional changes to create the right conditions for introducing the new technology (see also Kuznets, 1973).

Finally, Abramovitz argued that the more that backward countries contain redundant workers in farming and petty trade, the higher the opportunity for rapid growth by improving the allocation of labour, i.e. moving employment from primary to secondary sectors, using Colin Clark's terminology.

The work by Abramovitz hence represents a valuable link between the retardation theory and the convergence hypothesis. He stressed the relevance of idiosyncratic factors in shaping the convergence, proposing that convergence can occur only among countries characterized by the same social capabilities (Abramovitz, 1994). Among the qualifying features, the share of employment in the mature sector plays an important role, and the convergence appears to be shaped by the possibility of moving employment from one branch to another. This amounts to changing the employment share of economic branches of production.

2.4 An overview of the different analytical approaches

The interpretation of the relationships between industrial development and economic growth, as articulated by Kuznets, did not lie on a sound analytical model, but rather on a body of well articulated "appreciative theorizing" (Nelson, 1995; Syrquin, 2010). The formal implications of such an approach to economic growth were revealed only in the late 1960s. Indeed, the original growth models in the neoclassical tradition represented one-sector economies, in which the achievement of balanced growth is basically at odds with any possibility of uneven growth dynamics across sectors, and hence with the existence of a process of structural change.

The growth model most in line with the interpretation provided by the so-called three-sector hypothesis is that developed by Baumol (1967). This model leads to unbalanced growth in the transition phase. The economy consists of two sectors, one technologically stagnant, with only sporadic increases in productivity, and one technologically progressive. The former is closer to the idea of service sectors, the latter to that of manufacturing. In his model, Baumol focuses on labour as the most relevant input, assuming that outlays other than labour could be ignored. The output of the stagnant sectors grows as a function of employment levels, while that of the progressive sectors depends both on employment levels and on the rate of labour productivity:

$$Y_s = aL_s \qquad Y_p = bL_p e^{gt}$$

where the total labour is divided between the two sectors: $L = L_s + L_p$. Nominal hourly wages are the same in both sectors, and they grow in function of productivity growth in the progressive sector:

$$w = e^{gt}$$

It is straightforward that in the stagnant sector, unit labour costs grow unboundedly, while in the progressive sector it is an inverse function of the constant b:

$$c_s = \frac{wL_s}{Y_s} = \frac{e^{gt}}{a} \qquad c_p = \frac{wL_p}{Y_p} = \frac{1}{b}$$

If such conditions hold, one should expect the demand for the output produced by the stagnant sector to decline. By assuming that the price elasticity of demand for the two outputs is very close to unity, and that prices are proportionate to costs, the relative costs for the two commodities would remain constant:

$$\frac{c_s}{c_p} = \frac{wL_s}{wL_p} = \frac{L_s}{L_p} = k$$

And hence the output ratio between the two sectors will be given by the following relationship:

$$\frac{Y_s}{Y_p} = \frac{aL_s}{bL_p e^{gt}} = \frac{ka}{be^{gt}}$$

Such ratio declines constantly over time. However, it can happen that for some reason it is desirable to keep constant between the two sectors, despite the changes in the relative costs and hence in the relative prices:

$$\frac{b}{a}\frac{Y_s}{Y_p} = \frac{L_s}{L_p e^{gt}} = k$$

We can now derive the labour quantities in the two sectors:

$$L_s = k(L - L_s)e^{gt} = \frac{kLe^{gt}}{(1 + ke^{gt})} \qquad L_p = \frac{L - L_p}{ke^{gt}} = \frac{L}{(1 + ke^{gt})}$$

As t approaches to infinity, it is clear that the labour input in the progressive sector goes to zero, while that of the stagnant sector tends to absorb the whole labour force. If the output share of stagnant sector is allowed to increase, the transfer of labour force from the progressive sector is even greater. In an economy in which the output ratio between the two sectors is kept constant, the growth of total output tends to zero.

The analytical model has been complemented by extensive empirical evidence provided by the author (Baumol *et al.*, 1985, 1989), who showed how the differential rates of productivity growth of manufacturing and services are associated with a large-scale reallocation towards the tertiary sectors. More recently, models focused on the supply side like Baumol's have been proposed by Ngai and Pissarides (2007) and Acemoglu and Guerrieri (2008), who provide frameworks in which Baumol's results can arise endogenously from the combination of different capital intensities and capital deepening in the aggregate.

The three-sector hypothesis also has analytical counterparts lying on the demand side. Of these contributions, it is worth mentioning the models by Echevarria (1997), Laitner (2000) and Kongsamut *et al.* (2001), which all build upon the standard general equilibrium framework of neoclassical growth models, in which nonhomotetic preferences are integrated. These latter in turn are the main thing responsible for the changes in the sectoral composition of the economy, to which technological change is exogenous. At a more general level, the classical reference when speaking about demand-side models of structural change is undoubtedly Luigi Pasinetti. In his contributions, Pasinetti (1981, 1993) emphasized the relevance of demand dynamics, stressing that the earlier empirical investigation of demand dynamics could be dated back to 1850, when Ernst Engel studied the relationship between demand patterns and income change. Accordingly, the author proposed a view of the consumer as characterized by a hierarchy of needs, or order of priorities among groups of needs and services. Economic growth implies necessarily the growth of income. As income increases, consumption choices tend to shift from one group of goods and services to another. This shift of consumers across demand schedules is the main cause of structural change. As he put it, "employment in each sector *i* ... moves through time at a rate of change equal to the rate of population growth plus the rate of *increase of per capita demand* for commodity *i*" (Pasinetti, 1981, p. 95; emphasis added).

The main problem with such theoretical frameworks consists of the exogeneity of technological change, which however solved in the stream of models within the endogeneous growth theory, or Schumpeterian growth theory, approach which is mostly compatible with the existence of different sectors in the economy (Aghion and Howitt, 1992; Grossman and Helpman, 1991; Romer, 1990). However, such models, while having the merit of endogenizing technological change, provide a less convincing framework to the analysis of structural change, as they are based on the symmetry assumption across the different sectors. If the economy expands evenly across all sectors, which therefore all show the same growth rates, there is no room for structural change. This problem has been addressed by analytical models based on the replicator dynamics, which will be introduced in the following section.

2.5 Structural change and technological change: the complementarity between Kuznets and Schumpeter

We have seen that the foundations of the empirical analysis of structural change were laid down in the 1930s, within the context of the three-sector hypothesis. However, although innovation was explicitly seen as a crucial factor shaping the rate and direction of structural change, it has been often evoked as exogenously affecting the dynamics at stake. This is rather surprising, provided that Kuznets held in the 1960s the same chair at Harvard that Schumpeter held for about 20 years until 1950. And it was just in the seven years of his early career at Harvard that Schumpeter realized the work which is much closer to the analysis of structural change, i.e. the *Business Cycles* (1939) (McCraw, 2006).

The same applies also to the analytical models emphasizing the importance of the demand side. Indeed, the articulated framework elaborated by Pasinetti showed the same key limitation, in that technological change is given exogenously without motivation and justification (Syrquin, 2010).

Interesting efforts to cope with such weaknesses can be found in Metcalfe *et al.* (2006), who establish a connection between Pasinetti's emphasis on demand dynamics and growth retardation through technological change. The basic element is the grafting of the contributions of Adam Smith and Allyn Young (1928) into a model of industrial growth, accounting for dynamics of productivity growth as induced by output growth, through the self-propelling mechanisms fed by innovation activities. On the supply side, Metcalfe (2003) develops a replicator model able to account for the dynamics of industrial retardation as articulated by Kuznets and Burns, who are indicated therein as clear predecessor of the evolutionary approach to economics (Nelson and Winter, 1982).

Interestingly enough, the evolutionary approach is grounded on Schumpeter's contribution, and despite the several elements of complementarity between the three-sector hypothesis and evolutionary thinking, the acknowledgement of such interactions has long been neglected by the evolutionary scholars. All in all, we can notice that the intertwining of Schumpeterian and Kuznetsian dynamics has been successfully articulated, although not explicitly or maybe unconsciously. This is even more evident in the empirical analyses put forward by the "neo-Schumpeterian" authors who animated the scientific activity at the Science Policy and Research Unit (SPRU) of the University of Sussex, with contributions by Carlota Perez, Chris Freeman, Luc Soete and Giovanni Dosi, all of whom provide long-run interpretations of the interactive dynamics of technological and structural change, which emphasize the role of technologies in the change in the industrial composition of advanced economies, like in the case of the transition to the information economy, as well as the S-shaped process of industrial evolution. However, references to Simon Kuznets or Arthur Burns can hardly be found in their works (Perez, 1985, 1987; Freeman and Soete, 1990; Dosi, 1982).

In our opinion, the analysis of the interrelationship between technological change and structural change provides a different, and yet complementary, perspective when undertaken from a supply-side point of view. It allows for understanding the reciprocal influences, whereby structural change is likely to shape the rate and direction of technological change, and vice versa.

The interplay between Schumpeterian dynamics and retardation theory can be far-reaching and enhance the understanding of differences in the transition dynamics typical of structural change processes (Quatraro, 2009a, 2009b). Schumpeter indeed argues that innovation represents the main engine of economic progress within the capitalistic system (Schumpeter, 1928, 1939). Moreover, such an engine is constantly switched on, as "the opening up of new markets, foreign or domestic, and the organizational development ... illustrate the same process of industrial mutation ... that incessantly revolutionizes the economic structure from within, incessantly destroying the old one, incessantly creating a new one" (Schumpeter, 1942, p. 83).

Kuznets himself stressed the bearing of Schumpeter's approach upon the analysis of structural change. He noted that the process of creative destruction entails two parts, the creation of new combinations on the one hand, and the destruction of the old ones on the other. The introduction of radical innovations alters the structure of the economy, creating new jobs and making the existing ones obsolete. This in turn engenders a dislocating effect upon employment, which tends to shift from the old sector to the new one, with major difficulties in terms of switching costs (Kuznets, 1972).

Economic agents operate in environments shaped by the conditioning influence of factors both internal and external to the economic system. When there is an unexpected change in one or more of these factors, economic agents have to adjust. The way this happens may reside either within the comfortable borders of the existing practice, or outside its range. Creative response is an adaptation effort carried out by doing something completely new, which alters the data of the system (Schumpeter, 1939, 1947).

Innovation emerges out of the process of competition within the capitalist system, as an outcome of the creative response. Economic performances and innovation performances are characterized by complementary cycles. Innovating activities appear to be clustered in time, long after the expanding stages of the industry. Such a lag is due to a delayed diffusion of entrepreneurial ability among firms within the sector (Schumpeter, 1939).

The bringing about of innovation is a specific task of the entrepreneur, who is the actor getting things done by bearing the risk of putting resources to untried uses (Schumpeter, 1911, 1928). The scope for profiting from innovation is what pushes the entrepreneur to choose to creatively react rather than passively adapt. These profits, however, are not indefinitely available in the industry, but are instead temporary. The competing down process is likely to deter further innovation efforts (Schumpeter, 1939, 1942). The decision to innovate holds as long as the benefits are larger than the costs. When a saturation level is reached, in which the expansion on the supply side goes faster than that on the demand side, innovation efforts are likely to gradually fade out.

While Schumpeter's analysis of the cyclical behaviour of economic and innovation activities received major criticisms, mainly concerning his methodology, it had the merit of drawing attention to the role of innovation in the process of structural change (Kuznets, 1940). In particular, in his 1939 work Schumpeter focused on three countries, showing that the process of economic development was led by five industries and three institutional innovations. Thus in his work is found the concept of "leading sector", which was common to other authors in the same years, such as Kuznets and Burns (Rostow, 1975).

Schumpeter's and Kuznets' work turns out therefore to be strictly complementary. In the former indeed there is scarce attention to the dynamics of structural change, which are the main preoccupation of the latter. The change in the economic structure, in the sense of a change in the allocation of employment across different industries, is likely to shape and eventually rejuvenate the dynamics of productivity growth. Within each industry the process of Schumpeterian competition is likely to shape the dynamics of innovating behaviour.

A sequence between creative reaction and creative destruction can be detected. Firms within the established sector begin to innovate as soon as the room for further expansion gets smaller. Firms innovate to adjust to changes in the environment they operate in, so as to preserve or to gain further market shares. Innovation becomes systematic as opposed to sporadic: a local innovation system emerges, where relevant knowledge externalities become available and firms rely upon the introduction of innovation as a source of competitive advantage. When the number of innovating firms increases but the productivity growth rate within the industry keeps on reducing, the boosting effect upon innovation disappears, and innovation efforts are then directed outside. Creative destruction emerges as the force creating a new structure to the detriment of the old one.

Growth rates are unevenly distributed not only across industries, but within the same industry they are unevenly distributed across different regions. Thus one would expect the process of economic growth to be driven by different sectors in different regions. By the same token, one would also observe different kinds of innovation dynamics within each region, according to the relative evolution of the economic structure.

The feedbacks between retardation of growth rates and Schumpeterian competition thus give rise to a self-propelling process featured by endless economic change, as shown in Figure 2.1.

2.6 Conclusions

This chapter has developed an ideal path from the original analyses of structural change to the combination of such an approach with a proper account of technological change. Both Kuznets and Schumpeter place importance on the concept

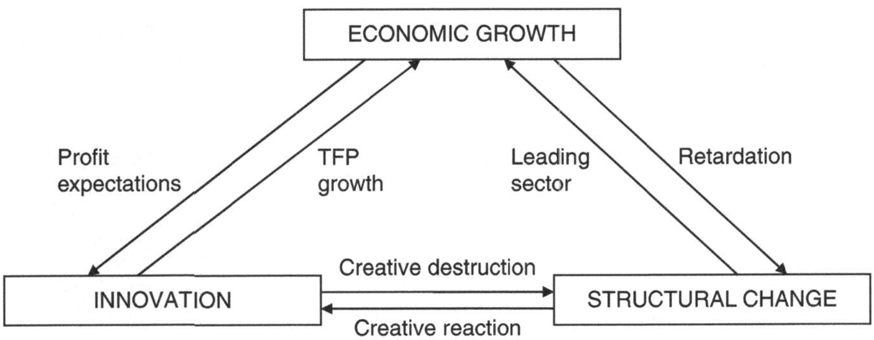

Figure 2.1 Feedbacks among innovation, structural change and economic growth.

of leading sector, and emphasize the mechanisms by which countries take the lead in international competition on the basis of their industrial and, strictly related, technological specialization. Creative destruction, enabled by innovation efforts, is a key part of the process leading to the emergence of new sectors or the rejuvenation of established ones. Innovation efforts are in turn the outcome of a creative response mechanism set in motion by unexpected changes in the economic environment, out of which structural change represents an important part. Thus structural change and technological change, in this perspective, affect each other in a dynamics of mutual dependence. Economic agents, however, would have no reason to innovate but to protect the prospects for profits that can be jeopardized by the changes in the economic environment. In this direction, structural change becomes both an incentive and an outcome of an endogenous process of technological change. The linkages between the two aspects deserve therefore to be fully articulated in a more coherent framework. For this reason in the next chapter we will elaborate upon the concept of technological knowledge and on its representation in the economic literature, so as to reach an approach better suited to being integrated into the analysis of structural change.

Notes

1 For the sake of completeness, Kuznets' view of structural change was even broader. He actually emphasized the necessary changes in the *social and institutional structure*, which are strongly related with changes in economic structure, and which create the conditions to implement technological innovations once they are introduced in the system.

2 It is worth clarifying that Metcalfe speaks about "relative" rather than "absolute" retardation.

3 The economics of technological knowledge

3.1 Introduction

The economics of knowledge is a discipline which has been developing rapidly in recent decades. Obviously, the importance of creativity for the production of goods and wealth is not a recent discovery within economics. The earlier treatment can indeed be found already in Adam Smith's first four books of the *Wealth of Nations*. After more than a century, Alfred Marshall elaborated upon Smith's contribution by proposing a formal systemic account of the role of knowledge in the production process. In particular, Marshall made it very clear both in *Industry and Trade* (1919) and in the *Principles of Economics* (1890 [1920]) that knowledge is a key input in the production process and the main engine of economic growth.

A clear step forward has been marked by scholars like Simon, Hayek and Fritz Machlup. Of Simon's contributions, particularly important is the analysis of the role of memorization in learning processes as well as the economic analysis of generation and transmission of information (Simon, 1982). Hayek (1945) introduced a key concept which is all the more relevant in today's theoretical approaches, concerning the fragmentation and dispersion of knowledge across the economic agents. Machlup (1971) proposes instead one of the former systematic accounts of the mechanisms by which knowledge is produced, diffused and exploited in the US from an economic viewpoint. Naturally, such precursors of an economics of knowledge are featured by major limitations in that they tend to use interchangeably the terms "information" and "knowledge", and therefore are prone to define the domain of an economics of knowledge in a fairly broad way. As suggested by Steinmuller (2002), such confusion has led the development of economics of knowledge to neglect important aspects for the field, like learning and cognition. This is also due to the implicit assignment of the activity of knowledge production to a separate sector of the economy, which is unlikely to communicate and exchange with the functions related to the production process of goods. Such traditional approach privileged the idea of an "off-line" production of knowledge, neglecting the importance of the "on-line" dynamics in which learning and interactions are central. Before proceeding, it may be therefore useful to clarify the distinction between knowledge and information.

By the latter one essentially means data and concrete facts, which are independent of any interpretation effort. Knowledge is instead a particular mental representation of information, within a specific context of interpretation (Nelson and Winter, 1982; Arora and Gambardella, 1994). The production of technological knowledge emerges as the result of a cumulative process, shaped by vertical and horizontal indivisibilities. The different faces of learning are primary sources of new knowledge, which displays a high degree of embeddedness with the context of activity. In view of this, firms' search process, far from being random, is likely to be bounded by a multidimensional corridor which characterizes the localized nature of technological knowledge. The effective access to external knowledge is constrained by the absorptive capacity firms actually possess, due to the relevance of knowledge complementarities. It follows that time and space in which the search process starts do matter, since it can shape the subsequent activity of exploration through the definition of a space in which it is more likely the firm will proceed (Antonelli, 2001, 2005; Dosi, 1988; Teece, 1988).

In this perspective the definition of economics of knowledge as a discipline is related to the analysis of the institutions, technologies and social regulations that can facilitate the efficient production and use of knowledge (Foray, 2004). The economics of knowledge should therefore shed light on the incentive schemes for economic agents to the allocation of resources to the production of knowledge, on the dynamics of socialization and disclosure, as well as on the conditions favouring the effective exploitation of knowledge available in the economic environment.

These issues are clearly influenced by the features of knowledge as an economic good, and are likely to stimulate the articulation of specific models of knowledge production and reproduction, as well as specific analytical representations of knowledge for theoretical and empirical assessments. In this chapter we develop a historical discussion of the concept of knowledge as an economic good, as well as the linkages with modes of production and operational translations (Krafft and Quatraro, 2011). Without pretending to be exhaustive, we propose a path moving from the notion of knowledge as a public good to the collective knowledge approach, which provides the bases to develop a structuralist conceptualization of knowledge able to be integrated into the analysis of structural change within a broader framework featured by complex system dynamics.

3.2 Knowledge as an economic good

One of the key issues related to the development of an economic analysis of the creation and exploitation of technological knowledge is its characterization as an economic good. This indeed has important consequences for the identification of the set of economic incentives to the creation of knowledge and therefore for the institutional setting favouring the generation, diffusion and exploitation of technological knowledge.

In this respect, the pioneer contributions by Kenneth Arrow represent the former attempt to elaborate an analytical model discussing the resource allocation to knowledge production. Arrow's (1962) approach is based on the consideration of knowledge as a *public good*, i.e. characterized by high levels of indivisibility, non-excludability, non-exhaustibility, non-appropriability and non-rivalry. In this view knowledge is therefore a good that is difficult to control privately, so as to prevent other persons from its utilization: as soon as knowledge is disclosed, it slips out of one's grasp. Moreover, the fact that many people use the same knowledge does not affect its value. That is to say, over-utilization is not likely to spoil knowledge effectiveness. Related to this, the same "piece" of knowledge can be used by one person with no limitation for simultaneous use by any other person. While the same pair of shoes can hardly be used by two persons at the same time, the same knowledge, say the same theorem, can be used by a potentially infinite number of persons simultaneously. For these reasons, the benefits stemming from the production of knowledge are not appropriable, and hence its tradability through the traditional market mechanisms is not viable. This implies that the market is not able to provide the appropriate incentives to the production of socially desirable amount of knowledge. In particular a trade-off between private and public incentives arise, such that economic agents would be willing to commit a lower amount of resources to knowledge production than that necessary to maximize the benefits for the society. In other words, the functioning of the market leads to suboptimal resource allocation to knowledge production. For this reason, the public provision of technological and, especially, scientific knowledge, represents in this framework as a basic remedy to under-provision. This has led to the actual implementation of knowledge commons and to the revival of endorsement and support to universities and public research centres (Arrow, 1962; Nelson, 1959).

A major shift in the economic analysis of technological knowledge took place when the established characterization of knowledge as a public good was challenged by an approach emphasizing the quasi-private aspects grounded on high levels of natural appropriability and exclusivity (Nelson and Winter, 1982). The key distinction between off-line and on-line processes of knowledge production becomes particularly relevant in this respect (Foray, 2004). The former usually refer to formal research and development activities separated by the production process, while the latter refer to the acquisition of new technological knowledge by means of learning dynamics. The notion of learning-by-doing and learning-by-using brings about the reversal of the top-down approach typical of the supporters of the public good argument, so as to propose a bottom-up mechanism in which knowledge is a sort of by-product of the production process, and as such is highly idiosyncratic to the context of production. The relationship between producers and users also becomes very important in that the latter can provide useful knowledge on how to improve new products just placed on the market, so as the eliminate possible drawbacks inherent to new designs. Learning-by-using is therefore another dimension of the learning process which contributes the accumulation of technological knowledge (von Hippel, 1988; Rosenberg, 1982).

Knowledge produced this way is essentially tacit, i.e. it is embedded in the competences and skills developed by economic agents in the course of daily execution of production routines. The concept of tacit knowledge, as is well known, has been put forward by Michael Polanyi (1958, 1966) to indicate a form of knowledge distinct from the knowledge explicit in conscious cognitive processes, and yet strictly complementary to them. The dictum that one person knows more than he can tell has been subsequently rapidly absorbed by scholars dealing with the analysis of technological change in the fields of economics and management. Nelson and Winter (1982) emphasize the importance of the tacit dimension of knowledge produced through "on-line" dynamics:

> the knowledge that underlies skilful performance is in large measure tacit knowledge, in the sense that the performer is not fully aware of the details of the performance and finds it difficult or impossible to articulate a full account of those details.
>
> (Nelson and Winter, 1982, p. 73)

It is worth stressing that a body of knowledge does not appear as tacit per se. It can be more tacit for some persons than others. Moreover, tacit knowledge can be codified, although with significant costs and effort. All in all, tacit knowledge can hardly be fully codified, though once codified, tacit knowledge on the "codebook" to interpret it can be necessary (Cowan and Foray, 1997; Cowan *et al.*, 2000; Foray, 2004). A major distinction between articulated and unarticulated knowledge can be therefore introduced. There may be knowledge that is potentially codifiable, but whose codification requires an effort that is not profitable. In this direction "knowledge is codified (sometime, somewhere) but not articulated (now, here)" (Cowan *et al.*, 2000, p. 229). Information flows imply codification and decodification efforts, and hence the issue of intelligibility. Unintelligibility may derive not only from differences in the natural language, but also from differences in its use. We can grammatically understand someone's language, but we cannot understand the real message content because of the inability to grasp the set of norms ruling language use (Hymes, 1972). Different kinds of tacit knowledge can be thus defined, according to different awareness levels, which are very relevant in investigating linguistic and semiotic determinants of new knowledge creation. Such an approach gives new strength to the concept of absorptive capacity (Cohen and Levinthal, 1990). The way people codify their own knowledge gains relevance in this perspective. The code they use to pack and unpack knowledge matters in assessing the success likelihood of a specific process of knowledge exchange (Amesse and Cohendet, 2001).

Insofar as knowledge stemming from learning dynamics is essentially tacit for persons outside the context of production, the risk of unintentional leak is highly reduced. Knowledge now is transferred only if the agents undertake the effort to do so. In other words, knowledge is sticky (von Hippel, 1994), so that the extent for knowledge spillovers appears to be limited, and in any case it is far from automatic. Knowledge externalities are not "in the air", and economic agents may appropriate at least partially the benefits from knowledge production. In such a context, there are clear economic incentives to allocate also private resources to the production of knowledge. The implementation of an effective institutional setting able to assign and enforce property rights on produced knowledge enhances the dynamics of knowledge as a proprietary good, providing the basis for an efficient use of markets for the exchange of knowledge (Arora *et al*., 2001).

The two "paradigms" discussed so far imply two different solutions to the trade-off between individually and socially optimal allocation of resources to the production of technological knowledge. Public procurement and public subsidies are on the one hand the most important institutional tool to foster the production of knowledge when it has mostly the properties of a public good. The creation of conditions for effective tradability of knowledge on markets, like the strengthening of patents and copyrights systems, represents the best solution when the proprietary aspects of knowledge are more pronounced. In both cases a clear trade-off between static and dynamic efficiency takes place, according to which the creation of the conditions to enhance the production of technological knowledge limits the functioning of the market economy either by imposing temporary monopoly power through the patent system or by endorsing the intervention of the government to address the market failures.

Besides these two alternative approaches to the economic analysis of technological knowledge, a new one recently emerged based upon the renewed appreciation of the role of external knowledge as an essential input in the production process of new knowledge. The collective knowledge argument is very promising in that it provides an important basis to the development of a structuralist approach to technological knowledge and therefore to the articulation of the relationships between knowledge and economic structure. Before examining the matter, however, it is important to stress that different economic characterizations of technological knowledge imply different views on the dynamics of the knowledge generation process as well as on the analytical representation of knowledge for empirical assessments. In the next section we will outline the implied consequences for what concerns the "public good" and the "proprietary good" frameworks, so as to better appreciate the important step forward represented by the collective good idea.

3.3 Modes of knowledge production and analytical representations

Besides the evolution of our understanding of knowledge as an economic good, the history of economics of knowledge has been marked by different conceptualizations of the process by which knowledge is generated and exploited, as well as different ways to translate it into an operational notion suitable of empirical assessment. The co-evolution of these aspects is characterized by implicit or explicit sets of relationships that deserve to be investigated at more depth.

3.3.1 Knowledge as public good, the linear model and the extended production function

The Arrovian approach, according to which knowledge shares mostly the properties typical of a public good, has clear implications in terms of governance of the process of knowledge creation, providing support in particular to the linear model of knowledge production, as well as to the modelling of knowledge as an exogenous factor affecting the dynamics of economic growth (Antonelli, 2005).

The former attempts at providing empirical accounts of the dynamics and the effects of innovation appeared only in the late 1950s. The studies by Griliches (1957) and Mansfield (1961) on the diffusion of innovation can be viewed as the earlier empirical efforts in this sense. However, very little was known at that time about knowledge and in particular about its production and exploitation. The earlier empirical works in which the word "knowledge" appeared to refer to a factor affecting the production of firms can be dated to the late 1970s. Zvi Griliches turned out to be a pioneer in the field again. In his 1979 paper indeed he proposed the famous extended production function, which paved the way to a wide body of empirical investigations. In this paper the traditional production function was extended so as to include an additional explanatory variable, as follows:

$$Y_i = C_i^{\alpha} L_i^{\beta} K_i^{\gamma} \tag{3.1}$$

Where C is the fixed capital stock, L stands for labour services and K is the knowledge capital used by firm i. Strangely enough, the empirical literature has generated a great deal of confusion on this contribution, as it is usually taken as key reference in papers using the so-called "knowledge production function" approach. We believe this is due to a basic misunderstanding. Indeed, Griliches in his article made some steps forward to give an empirical meaning to the K term. To this purpose he proposed the following relationship:

$$K = G[W(B)R, \upsilon] \tag{3.2}$$

where R is R&D expenditures and υ is a set of unobserved disturbances. The term $W(B)$ is instead a lag polynomial describing the relative contribution of past and present R&D expenditures to the accumulated level of knowledge. Clearly, this representation is one more application of the distributed lag literature, which influenced Griliches to a great extent. Far from proposing a knowledge production function, this relationship was simply the formalization of the concept of knowledge capital stock, which the author subsequently used in his 1980 paper on the US productivity slowdown (Griliches, 1980). In a nutshell, the 1979 paper offered the formal basis to the application of the permanent inventory method to calculate the knowledge stock starting from R&D expenditures, which are then considered as a flow measure.

The specification of knowledge capital also called for a proper account of the effects of knowledge spillovers, i.e. knowledge borrowed or stolen from other firms or industries that can equally affect productivity of the observed firm or industry. Knowledge spillovers have been accommodated in an extended production function at the firm level by including a proxy for the aggregate stock of knowledge available within the industry in which firm i operates:

$$Y_i = C_i^{\alpha} L_i^{\beta} K_i^{\gamma} K_a^{\mu} \tag{3.3}$$

Such equation enables us to distinguish between the total effect of aggregate private knowledge and the total spillover effect. Since all private knowledge is supposed to spill over to some extent, the total effect of all private knowledge at the aggregate level is given by $\gamma+\mu$ (Griliches, 1979, 1992).

On the basis of the argument elaborated so far, we may provide some insights about the possible theoretical underpinnings to the concept of knowledge capital stock. Indeed, we lack an explicit theoretical reasoning on technological knowledge leading to its operationalization in terms of knowledge capital stock. A quotation from Griliches (1967) may be of some help here:

> For example, let investments affect the level of patenting with a lag whose generating function is given by $W_1(z)$, let these new inventions be embodied in new investment with a lag $W_2(z)$ and let new investment affect total factor productivity with a lag $W_3(z)$; then the total lag distribution of productivity behind investment is given by $W(T) = W_1(z)W_2(z)W_3(z)$.
>
> (Griliches, 1967, p. 20)

It is clear that the application of lag generating functions to investments measures so as to get a stock implies an underlying sequential process that starts with R&D investments to yield a proxy of cumulated knowledge that in turn is supposed to show some effects on economic performances. In this direction, we believe it would not be so inappropriate to suggest that knowledge capital stock implies a vision of knowledge accumulation as an outcome of a linear process like this: science precedes technology development, which then comes to be adopted by firms, and finally affects production efficiency.

After all, Vannevar Bush's 1945 report to the US president had long been the main reference text for students of science and technology. Therefore it is likely that the articulation of the linear model he proposed has influenced the way scholars from other fields looked at technological knowledge as well. Moreover, Kline and Rosenberg's critique came only in the 1980s, as did many of the works that opened up a new view on knowledge and innovation providing the basis for the knowledge production function approach (Bush, 1945; Kline and Rosenberg, 1986; Balconi *et al.*, 2009).[1]

3.3.2 Knowledge as a proprietary good, knowledge interactions and the knowledge production function

The approach to knowledge as a proprietary good lends itself to a rethinking of the model of knowledge production, and therefore of the way knowledge is analysed in empirical settings. After all, the inclusion of knowledge capital stock within an extended production function approach allows economists to preserve the basic microeconomic assumptions about production sets out of which firms take their profit-maximizing choice. However, such approach still assumes the existence of a separate R&D sector, i.e. an off-line mode of production, that is partly responsible for the change in the production technology, and hence of the shift of the production function (Nelson, 1980).

Because of this limitation, such a representation began to be challenged mainly by evolutionary economists, who proposed expanding the view on technological knowledge so as to account for its inherent compositeness. At the same time, scholars of science and technology started criticizing the linear model, by proposing an alternative view basically drawing upon systemic models of innovation based upon the interaction among different and yet complementary institutions involved in the complex business of knowledge production (Kline and Rosenberg, 1986; Gibbons *et al.*, 1984).

A couple of Dick Nelson's contributions in the early 1980s provided a clear statement of the problems with the concept of knowledge capital stock, along with the theorization of a more articulated concept of knowledge, understood as a set of capabilities guiding the search processes undertaken by organizations performing R&D. Such capabilities may themselves be the outcome of R&D activities, and are likely to improve over time due to dynamic increasing returns stemming from learning-by-doing dynamics (Nelson, 1980, 1982).

In this sense, such contributions may be viewed as pioneering in the attempt at opening the black box of technological knowledge so as to explicitly improve upon Griliches' and Mansfield's former operationalizations. Moreover, they also proposed a more realistic view in which science and technology are far from sharply differentiated. There are a number of institutions producing knowledge, some of them are public while others are private, and it is not possible to identify a one-to-one mapping from science to public institutions or from applied technology to private business firms. Scholars must acknowledge that different kinds of organizations take part in the process of knowledge production, like firms, research labs and universities (Nelson, 1982, 1986).

This set of arguments has been well received mostly in the literature dealing with knowledge production at the aggregate level. In particular the literature on regional systems of innovation provided a fertile ground to develop the implications of this new view (Cooke, 1996; Cooke *et al.*, 1997). Regional economists translated the idea that knowledge is the result of the interaction of a number of complementary inputs provided by different research institutions into the concept of knowledge production function. The differences with the concept of knowledge capital stock are clear. Knowledge is no longer the mere result of cumulated

R&D spending subject to decreasing returns. The knowledge production function provides a mapping from knowledge inputs to knowledge outputs that appears as follows:

$$\log(K_t) = \alpha + \beta \log(R_t) + \gamma \log(U_t) + \delta \log(Z_t) + \varepsilon \tag{3.4}$$

where K stands for a measure of knowledge output, say patents, R stands for the industry R&D and U represents the university research, while Z includes a proxy for the concentration of a given type of activity (Acs *et al.*, 2002; Fritsch, 2002). Equation (3.4) represents a production function, the arguments of which enter a multiplicative relationship, and hence are seen as complementary rather than substitute. The coefficients are in turn the elasticities of knowledge output to knowledge inputs.

On a fairly similar ground, the localized technological change approach has stressed that the dynamics of knowledge production are characterized by the joint utilization of internal and external knowledge, both tacit and codified. Mechanisms of learning, socialization and recombination are considered as crucial in a context characterized by the production of knowledge by means of knowledge itself (Antonelli, 1999).

The knowledge production function approach represents an improvement both from the theoretical and the empirical viewpoint, with respect to the concept of knowledge capital stock. It allows gaining a better understanding of the interactive dynamics leading to the production of technological knowledge, by accounting for possible dynamic increasing returns stemming from learning dynamics as well as knowledge externalities. However, knowledge on the left side of the equation is still conceived as an homogeneous stock, and little is said about the intrinsic heterogeneity of knowledge base. In other words, such representation still lacks proper cognitive models of knowledge production.

3.4 Recombinant growth and complex knowledge

The development of the knowledge production approach inevitably leaves us with a basic question as to the micro-founded mechanisms underlying knowledge production. In this respect, the interest in the cognitive mechanisms leading to production of new technological knowledge has recently emerged in the field of economics of innovation. This strand of analysis has moved from key concepts put forward by Schumpeter (1912, 1942) and Usher (1954) to elaborate upon the models proposed within evolutionary economics (Nelson and Winter, 1982).

In his seminal works, Schumpeter proposed viewing innovation as the outcome of a recombination process. Most innovations brought about in the economic system stem from the combinations of existing elements in new and previously untried ways. Such innovations appear to be mainly incremental. Radical innovations stem instead from the combination of existing components with brand new ones.

The contributions by Weitzman (1996, 1998) represent the former, and very impressive, attempt to draw upon such assumptions. His recombinant growth approach provides a sophisticated analytical framework grafting a micro-founded theory of knowledge production within an endogenous growth model. The production of knowledge is seen as the outcome of an intentional effort aimed at reconfiguring existing within a genuine cumulative perspective. However, there is no particular focus on the constraints that the combination of different ideas may represent, especially when these ideas are technologically distant. The only limiting factor seems to be the bounded processing capacity of economic agents.

The recombinant knowledge approach is based on the following assumptions. The creation of new knowledge is represented as a search process across a set of alternative components that can be combined with one another. However, within this framework a crucial role is played by the cognitive mechanisms underlying the search process aimed at exploring the knowledge space so as to identify the pieces that might possibly be combined together. The set of potentially combinable pieces turns out to be a subset of the whole knowledge space. Search is supposed to be local rather than global, while the degree of localness appears to be the outcome of cognitive, social and technological influences. The ability to engage in a search process within spaces that are distant from the original starting point is likely to generate breakthroughs stemming from the combination of brand new components (Nightingale, 1998; Fleming, 2001).

Incidentally, such an approach also enables us to better qualify the distinction between exploration and exploitation formerly articulated by March (1991). Most of the research in organization studies has usually seen search processes as ranging between two poles of a one-dimensional continuum, i.e. exploration and exploitation. The view of knowledge as an outcome of a recombination activity allows the introduction of two nested dimensions, defined according to the degree to which agents decide to rely either on exploration or exploitation or on a combination of both. To this purpose concepts like search depth and search scope have been introduced. The former refers to the degree to which agents intend to draw upon their prior knowledge, while the latter refers to the degree to which agents intend to rely on the exploration of new areas in the knowledge space (Katila and Ahuja, 2002).

Recombination occurs only after agents have put much effort into searching within the knowledge space. This strand of literature posits that knowledge so obtained is complex, meaning that it comprises many elements that interact richly (Simon, 1966; Kauffman, 1993). This has paved the way to an increasing number of empirical works based on the NK model proposed by Kauffman, according to which the search process is conducted across a rugged landscape where pieces of knowledge are located and which provides the context within which technologies interact.

The bulk of the focus is on the concept of interdependence among the pieces that are combined, while complexity is defined as the relationship between the number of components and the degree of interdependence (Fleming and

Sorenson, 2001; Sorenson *et al.*, 2006). Following the intuition on the importance of patent citations contained in the seminal paper by Manuel Trajtenberg (1990), the empirical implementation of the interdependence concept is based on the deployment of the information contained in patent documents, i.e. technological classes and citations to other patents. In particular, interdependence is considered as a powerful explanatory variable building upon the technological classes the patent is assigned to. The interdependence of a patent l is obtained in two steps. First of all one has to calculate the ease of recombination for each subclass $i\,(E_i)$, defined as the count of subclasses $j \neq i$ previously combined with class i weighted by total number of patents assigned to class i:

$$E_i = \frac{\sum j \neq i}{\sum l_i} \tag{3.5}$$

Then one can calculate the degree of interdependence of patent $l\,(K_l)$ by inverting its average ease of recombination:

$$K_l = \frac{\sum i \in l}{\sum_{i \in l} E_i} \tag{3.6}$$

This empirical approach allows for evaluating the relative probability of recombination of each technological class observed in the patent sample, and then to assign an average recombination score to a patent. The basic idea is that the more combinable the classes contained within a patent, the lower the degree of interdependence, as the technology is susceptible to being developed in a larger number of directions. On the contrary, should the classes be hardly combinable, then a relatively low number of possible combinations is possible, for which the technology turns out to show a high degree of interdependence. Such measure of interdependence is in turn expected to explain differentials in usefulness of inventions as proxied by the flow of citations received by patents over time.

Such framework clearly has the merit to push the economic discussion about technological knowledge beyond the conventional vision considering it as a sort of black box. It sheds light on the possibility to further qualify knowledge as proxied by patents, by better exploiting the information contained in patent documents. Moreover, it provides a former and innovative link between knowledge and complexity.

However, the notion of complexity used therein seems to be constrained to a generic definition of an object the elements of which are characterized by a high degree of interaction. As an implication the empirical effort does not go beyond the count of classes and of patents assigned to classes. The NK models fail to identify knowledge as an emergent property of an adaptive complex system, characterized by an architecture that can influence the actions at the micro and meso levels as well as being influenced as a result of what happens at lower layers. This requires first to make explicit a concept of knowledge structure and then to explore the different tools made available by different methodological approaches.

3.5 Conclusions

In the recombinant knowledge approach, technological knowledge is proposed to emerge out of a search process conducted across a knowledge space within which smaller units of knowledge are distributed. This search is aimed at identifying bits of knowledge that may be combined so as to generate new technological knowledge (Weitzman, 1996, 1998; Fleming, 2001; Fleming and Sorenson, 2001; Sorenson *et al.*, 2006).

While search may in principle be conducted across any area of the knowledge space, the set of competences possessed by economic agents, as well as the set of social and technological influences within which they operate, are likely to constrain their recombination activity to a well defined area of the knowledge space, thus providing some boundaries to evolutionary paths. In these conditions the search process is likely to be more effective when conducted on a local rather than on a global scale. As a consequence, the degree of localness matters in shaping knowledge production, and it makes relevant the extent to which combinable elements are complementary and similar to one another. In the meantime, the ability to engage in a search process within spaces that are distant from the original starting point is likely to generate breakthroughs stemming from the combination of brand new components (Nightingale, 1998; Fleming, 2001).

Although we agree that the recombinant knowledge literature made an important contribution, we do not assume that all new knowledge is created by the recombination of pre-existing one. The representation of knowledge we propose in the next chapter can encompass both the creative acts giving rise to entirely new observables and those which suggest new connections of observables already existing at a given time.

The next chapter will be focused on the elaboration of knowledge as an "organization" characterized by an evolving network structure. We will adopt a complexity-based perspective, according to which knowledge represents a subsystem of a wider hierarchy of nested structures. Both dynamic interactions within and between subsystems matter in generating a chain of interconnected emergence processes. Knowledge and economic structures will therefore manifest their strong mutual interdependence.

Note

1 We do not intend to go into the debate on the virtues and drawbacks of the linear model. The work by Balconi *et al.* (2009) provides an excellent synthesis in this direction.

4 Structural change and knowledge structure

An integrated framework

4.1 Introduction

In the previous chapter we have outlined the evolution of the concept of knowledge as an economic good, emphasizing the relationships of the different characterizations of technological knowledge with the concept of the knowledge generation process proposed by scholars of innovation as well as with the operational translation of knowledge for empirical assessment.

We have noticed that over time scholars of economics and management dealing with the study of innovation have refined the conceptualization of technological knowledge, providing more and more complex and articulated pictures. The establishment of the recombinant approach marks undoubtedly a step forward in the understanding of the cognitive dimensions of knowledge production, as well as in the search for a more plausible empirical representation of knowledge. Most importantly, such approach lends itself to the development of an interpretative framework grounded on complexity theory. In other words, the domain of inventions is viewed as complex systems in which inventions are highly interdependent. In this direction citation patterns are used as a starting point to derive indicators related to the NK representation of complex landscapes (Kauffman, 1993; Fleming and Sorenson, 2001; Fleming *et al.*, 2007).

However, the recombinant approach shows some important limitations that make difficult its use in a dynamic perspective. First, the recombination is supposed to occur among existing elements, so that the introduction of brand new elements in the technology landscape is not properly accounted for. Second, the architecture of the relations among the elements of the system is mostly static in their analysis, while architectural change is of paramount importance in complex designs (Henderson and Clark, 1990; Murmann and Frenken, 2006). Finally, the emphasis on the components of the technology limits their framework to the analysis of the artefacts, without extending the domain to the appreciation also of the agents' side of the "agents/artefacts" space (Lane *et al.*, 2009; Lane, 2011).

This chapter is meant to propose a framework to unify the agents' and the artefacts' domains, so as to represent the creation of knowledge as an emergent phenomenon generated by complex feedbacks between the two domains and between the elements within each of these systems. In this direction, the relationships between the changes in the economic systems will be further articulated. A crucial step to move towards a dynamic complexity-based representation is the grafting into the picture of an approach to knowledge as a collective good. This indeed allows for appreciating the dynamics occurring at the agents' levels and therefore provides the basis for a mapping between the network of innovating agents and the recombinant dynamics of technological knowledge. The next section will propose an outline of the collective knowledge approach. We then elaborate upon the concept of "structure" in order to give new lifeblood to the structuralist approach in economics, which is likely to provide a heuristic toolkit able to accommodate a dynamic view of complex system dynamics at different levels. We propose to understand the agents'/artefacts' space as a broad system made of subsystems which represent highly interdependent components. These subsystems are in turn made of lower-level interdependent subsystems, and so on and so forth. In other words, we will end up with a conceptualization of the socio-economic system as a hierarchy of nested subsystems characterized by modularity and recursivity (Arthur, 2009). In this context, changes in a subsystem are likely to affect not only the architecture of the subsystem, but also the architecture of the higher-level system as well as the architecture of the other subsystems showing a stronger interdependence with it. The relationship between economic structure and knowledge structure will appear therefore as only one part of a broader chain of feedbacks and adaptations in a constant tension to self-organization which is never fully accomplished.

4.2 Collective knowledge and interactive dynamics

The appreciation of the key role of external knowledge in the production process of new knowledge represents a far-reaching intuition for the analysis of the economic issues related to technological knowledge.

In view of this, the generation of technological knowledge can be depicted as an outcome of a collective undertaking strongly influenced by the availability of local sources of knowledge and by the quality of interactions (Allen, 1983; von Hippel, 1988). Specifically, technological knowledge, as it is used and generated by firms, stems from the combination of two basic inputs, i.e. internal and external knowledge. Technological knowledge is produced by firms operating within local contexts featured by the presence of a wide array of complementary knowledge sources, like other firms, universities and research institutions. Internal and external knowledge represent two strongly complementary inputs, hence featured by a very low or null degree of substitutability. As a consequence, neither of the two inputs may fall below a certain threshold without harming the knowledge production process (Antonelli, 1999).

The access conditions to external knowledge play a key role in the generation of new knowledge. The implementation of screening processes and absorption strategies by firms is the necessary condition to access existing external knowledge. Firms here need to undertake specific activities and efforts to integrate such external knowledge, which can be very different from those already possessed, into their internal knowledge production processes. In other words, access to external knowledge is harmed by the efforts agents must face to screen the markets of technological knowledge, and then acquire the relevant portion of knowledge produced and sourced externally (Pisano, 1996; Agrawal *et al.*, 2006; Beugelsdijk, 2007; Patrucco, 2009).

The localized production and diffusion of technological knowledge is the result of the collective strategies between firms' acquisition of external knowledge originated in both firms (e.g. suppliers, clients, rivals) and institutions (e.g. universities, R&D labs, TTOs), which are fostered by the presence of multiple, formal interactions and the active, intentional participation of firm in such knowledge exchanges (Dicken and Malmberg, 2001; Nicholas, 2009).

The analysis of the mechanisms through which knowledge results as a collective undertaking places a new emphasis on the role of interactions for the working of the markets for knowledge. The crucial analytical achievement of the research on the markets for knowledge is the appreciation that contractual devices and geographical proximity reduce the price of trading and exchanging of bodies of knowledge in the marketplace between the players of reiterated interactions (Arora *et al.*, 2001). Hence, geographical proximity complements and actually makes possible the markets for knowledge and the flows of transactions between, for instance, manufacturing firms, academic laboratories, new technology-based firms, consultants and knowledge-intensive services (Zucker, Darby and Armstrong, 1998; Zucker, Darby and Brewer, 1998; Boschma, 2005; Nakamura and Odagiri, 2005). It reinforces the effect of market interactions and leads to more effective knowledge exchanges: knowledge outsourcing and knowledge transactions may greatly benefit from agglomeration effects in that proximity facilitates the building of mutual trust and reciprocity that enables repeated interactions between co-localized firms (Feser, 2002; Gossling, 2003). The interplay between the role of proximity and the use of markets in a dynamic perspective allows us to appreciate the contribution of pecuniary knowledge externalities, as distinct from "untraded" knowledge externalities, to the effective exploitation of knowledge pools (Antonelli *et al.*, 2011).

Besides the interactive dynamics of knowledge production, the access to knowledge produced elsewhere in the economic system provides essential inputs able to make possible the exploitation of increasing returns generated by vertical and horizontal indivisibilities. Collective knowledge encompasses knowledge interactions, but is not limited to them. An agent can search across the knowledge landscape independently of the agents with whom they may interact. Obviously, the establishment of persistent networks for the exchange of knowledge is likely to increase the effectiveness of the search process and to avoid duplication costs. The ability to generate knowledge is embedded in a network of qualified relations. The notion of collective process marks therefore an important difference both with respect to the Arrovian tradition of knowledge as a public good and the approach to knowledge as a quasi-private good. Collective processes in fact are characterized by the role of the intentional effort, participation and contribution of each agent (Antonelli, 2005). Collective knowledge, in other words, is a shared activity that can be implemented only by interactive agents that belong to a community of practice and understanding. In this perspective knowledge can be represented as an activity, rather than a good, which is shaped by the commitment of resources to the access, absorption and implementation of external inputs, and by the advantages of increasing returns enabled by the exploitation of potential complementarities (Buchanan, 1965; Allen, 1983; Foray, 2004).

The collective knowledge approach implies therefore the existence of agents characterized by bounded rationality, which cannot have the full command of the whole knowledge space, and therefore need to access knowledge dispersed and fragmented in the economic system in order to feed the combinatorial dynamics leading to the production of new knowledge (Hayek, 1945). The network of innovating agents, its properties and its constituting elements, is a key structure for knowledge creation. The production of knowledge can consistently be represented by introducing two main general properties of knowledge, those of being (a) a co-relational structure and (b) a retrieval–interpretative structure (Saviotti, 2004, 2007). According to (b) the probability for any human being or organization to learn new knowledge falls with the dissimilarity, or distance, between the knowledge previously held and the external knowledge to be learned. According to (a) knowledge establishes generalizations by finding relations, or connections, between variables and concepts. The whole space of human knowledge can in principle be represented as a network the nodes of which are either variables or concepts and the links of which are the connections between different variables or concepts. Both the number of nodes and the number of links of such a knowledge network can be expected to change over time as new concepts and variables are discovered and new links are created between previously unconnected variables or concepts. The overall network of human knowledge can never be expected to be fully connected as the rate of addition of new nodes and that of creation of new links are unlikely to be identical at all times. Thus, the density or connectivity of the network of knowledge can be expected to fluctuate over time, rising or falling depending on whether the rate of creation of new links or the rate of creation of new nodes prevails. Such fluctuations are not in general random but are likely to be related to phases of a technological lifecycle or of a technological paradigm.

Such representation of knowledge, although it shares some similarities with the recombinant knowledge approach (Weitzmann, 1996, 1998; Olsson, 2000), goes a step further. According to this recombinant knowledge approach the creation of new knowledge is represented as a search process across a set of alternative components that can be combined with one another. A crucial role is played here by the cognitive mechanisms underlying the search process aimed at exploring the knowledge space so as to identify the pieces that might possibly be combined. The set of potentially combinable pieces turns out to be a subset of the whole knowledge space. Search is supposed to be local rather than global, while the degree of localness appears to be the outcome of cognitive, social and technological influences. The ability to engage in a search process within spaces that are distant from the original starting point is likely to generate breakthroughs stemming from the combination of brand new components (Nightingale, 1998). The network corresponding to the recombinant knowledge approach would have a constant number of nodes and a growing number of links. The step ahead with our approach is that the network of knowledge has a variable number of nodes and a variable number of links. Our approach encompasses the recombinant knowledge approach (creation of links between pre-existing nodes) but in addition allows the emergence of radically new concepts (introduction of new nodes).

4.3 A structuralist approach

The basic achievement that both the interaction among innovating agents and the knowledge stemming from such interactions can be represented as networks made of nodes and links opens up interesting perspectives on the creation of an interpretative framework based upon the concept of structure.

The development of structuralism in economics did not happen to be a lucky undertaking. The most remarkable efforts in this direction have been produced by the intellectual forces grouped around French journals like the *Revue d'Economie Politique*, the *Revue Economique* or *Economie Appliquée*, of which François Perroux was founding editor in 1944. Early reflections on the concept of economic structure can be found in Tinbergen (1952), Weiller (1935, 1952) and Perroux (1971). Since then, however, the discipline has moved far away from such an approach. Ragot (2003) articulated the idea that the economy is structured like a language, and provides an interpretation of the Walras equilibrium model as a clear exemplum of such an idea. In order to develop a structuralist approach to the analysis of knowledge and economy it is worth clarifying what is meant by "structure" and which "structuralism" we share.

The definition of the concept of structure can be articulated under different perspectives. Following Descombes (1980), we can move from the definition provided by mathematicians, and in particular the simplified and synthetic one by Bourbaki (1935):[1]

> We can now clarify what is to be understood, in general terms, by a *mathematical structure*. The feature common to the various notions ranged under this generic heading is that they all apply to sets of elements, the nature of which *is not specified*; in order to define a structure, one or more relations involving these elements may be taken … it may then be postulated that this or these relations fulfill certain conditions (to be enumerated), which are the *axioms* of the structure envisaged. To develop the axiomatic theory of a given structure is to deduce all the logical consequences of its axioms, *forbidding oneself any other hypothesis* concerning the elements under consideration (and especially any hypothesis with regard their particular "nature").
>
> (quoted in Descombes, 1980, p. 85)

Such definition has the advantage of escaping any formality providing the key ingredients of a structure: the elements and the connections among them. Moreover, it contains one major reason for critique of structuralism, that is the excessive focus on the relationships, forgetting the features of the individual elements. It has become clearer and clearer that the articulation of the concept of structure has been the most important pillar in the French intellectual sphere of the first half of the twentieth century. In the same environment, Jean Piaget, one of the key representatives of the structuralist school of thought, provided a definition of structure more suitable for utilization in social sciences:

> En première approximation, une structure est un système de transformation, qui comporte des lois en tant que systèmes (par opposition aux propriétés des éléments), et qui se conserve ou s'enrichit par le jeu même de ses transformations, sans que celles-ci aboutissent en dehors de ses frontières ou fasse appel à des éléments extérieurs. En un mot, une structure comprend ainsi les trois caractères de totalité, de transformation et d'autoréglage.
>
> (Piaget, 1968, pp. 6–7)

Once again, an inherent feature of the concept of structure is the focus on the whole, i.e. on the links between its constituting parts, rather than on their individual characteristics. It worth noting that Piaget's conceptualization places importance also on the self-organization as well as on the dynamics of the structure. This marks an important difference with the founders of the structuralist approach. This opens up the question as to what extent one can speak of one general structuralist approach, rather than of different kinds of structuralism. Traditionally, the founding text of the structuralism is the *Cours de Linguistique Générale* by Ferdinand de Saussurre (1916). The author makes therein a crucial dualism between the synchronic and diachronic dimensions characterizing social sciences, the former deserving to be properly investigated. The synchronic dimension concerns the symbolic identity of signs. The "langue" (as opposed to the "parole") is a collection of signs that have not a value in themselves, but only with respect to a system of relations with other signs. This system of relations represents the structure of the "langue" and its logical coherence is analysable independently of how signs are used and combined to form meaningful sentences. The use of signs occurs in historical time, while the structure of the system of signs is independent of time and represents the synchronic aspect of languages. The purpose of the scientist is to search for an invariant structure of relationships that may be used to analyse different systems of signs without the need of adaptation efforts. For this reason the structuralist approach is usually seen as essentially static and deterministic. Moreover, this search for an invariant structure of logical relationships is consistent with the concept of equilibrium: the "langue" is a system in equilibrium, in the same way as the economy in Walrasian theory (Ragot, 2003). In the same vein, Claude Levi-Strauss (1958, 1962) in his works show that the myths of "savage" tribes are characterized by a similar architecture referring to an invariant structure of binary oppositions, such that totemic beliefs apparently different can be compared and assimilated on the basis of basic antonymic trait pairs.

Théret (2003) proposes distinguishing between two broad approaches to structuralism, i.e. methodological and ideological structuralism. On other overlapping dimensions, structuralists can be put in three groups: the scientists supporting the methodological aspect, without any pretension to draw philosophical implications; the scientists implementing truly methodological approaches, but on the basis of ex-ante philosophical assumptions; and the ideological structuralists, who do not pay any attention to the scientific basis of their ideas. Levi-Strauss and Althusser can be assimilated to the last two groups, while Piaget is attributed to the second group.

Without pretending to dig into the debate on the different facets of structuralism, our attempt to recover structuralist thought is based on the development proposed by Piaget, which we think allows us to establish a link with the line of analysis which has grown under the label of complexity theory and which has much informed the analysis of economic phenomena.

Piaget in his work *Le structuralisme* (1968) articulates the distinction between structuralism as a philosophy and structuralism as a methodology, proposing that only the latter can be considered as an authentic approach, eminently epistemological. While he firmly defends such a view and applies it to psychology, the way he defines the concept of structure as well as its properties looks particularly interesting. According to Piaget, indeed, the structure represents a heuristic tool that is useful to understand the *dynamics* of human behaviour and its restless evolution. Such approach has also been labelled genetic structuralism (Théret, 2003), in that the main emphasis is not on the static properties of the structure, i.e. on the equilibrium of its architecture, but on the dialectic structure-behaviour, i.e. on the process leading to the generation of a particular structure with a given architecture. The structure, one could say, turns out to be an emergent property stemming from the dynamics of social systems. The structure cannot be understood without a proper account of its genesis and evolution. A structure emerges in the course of history as a result of a self-organizing process.

We can preliminarily conclude that the genetic structuralism paves the way for the development of an evolutionary approach to the analysis of structures, in which their architectures are not essentially stable, but change over time as an effect of mutations in the relations and in the elements. The adoption of a structure-based approach to knowledge turns out to be even more appropriate from a terminological viewpoint, as the idea of collectivism is much more related to a collection of agents rather than on the links among them. Structural holism is different from collective holism in that the whole is not just the juxtaposition of the individual elements, but is the outcome of the relationships among the components. The structuralism so conceived is consistent with the adoption of systemic thinking. A system can indeed be defined as a group of interdependent elements. Moreover, the idea of self-organization and emergence places structuralism very close to complexity theory. We could push the argument further by arguing that the analysis of complex system dynamics fully develops the potential of structuralist heuristic, by combining holism and individualism, i.e. combining the interest in the relationships with the attention to the properties of the single components of the system (Bloch and Metcalfe, 2011). The idea of nested hierarchies of architectures, moreover, pushes structures at the heart of the elements of which systems at each layer are made. The recursive principle implies that each architecture is a piece in higher-level architectures. The French language again can help the synthesis: *la structure devient structurante*. Before fully grasping the consequences of such argument, it is worth a quick detour around the concept of complex system dynamics, in order to develop later on the concept of knowledge structure as a complex design belonging to a system characterized by complex dynamics, through which changes in knowledge structure are amplified to other complex designs in the higher-order system.

4.4 Complexity and economics of innovation

The theory of complexity has become more and more popular in social sciences, and especially in economics, in recent decades. The research programme on complexity emerged in the context of post-Second World War developments in cybernetics, and in particular following the criticisms concerning the so-called first-order cybernetics (Bouraoui, 2009). The grafting of such line of reasoning onto the analysis of economic phenomena is as old as von Hayek "precocious play" on the epistemology of complexity (Hayek, 1967). Hayek's conception of complexity is still biased towards the "structural" aspects, neglecting the features of the individual elements of which the system consists. The degree of complexity can be defined as "the minimum number of elements of which an instance of the pattern must consist in order to exhibit all the characteristics attributes of the class of patterns in question". Complexity arises from the non-intelligibility of the whole pattern of interlinked behaviour. This makes point prediction unfeasible and draws the attention to ex-post relations between the emergence of new patterns and specific circumstances. Obviously, the higher minimum number of elements required, the more difficult it is to command the dynamics of interrelationships. Complexity is a basic property of spontaneous order, or self-organization, which emerges as "the result of human action, but not the execution of any human design" (Adam Ferguson, 1767; quoted in Hayek, 1973, Vol. I, pp. 20 and 150, note 19). The dispersed and fragmented character of knowledge (Hayek, 1937) plays a key role in the complex dynamics of spontaneous orders, in that it makes it impossible to fully command the whole, which is fairly more than the sum of the constituting parts, as well as the dynamics of its historical evolution. Self-organized systems for Hayek are indeed not only complex but also evolutionary (Bouraoui, 2009).

In Hayek's treatment can be found some important basic concepts for complexity theory as it is nowadays spelled out by scholars dealing with the economics of innovation and knowledge creation. In the same years, Herbert Simon (1969) told the now-famous story about the two clockmakers Tempus and Hora, and proposed defining a complex system as a system composed of interdependent elements. The modular and hierarchical structure of such kinds of system is then emphasized by introducing the concept of near-decomposability, according to which in modular systems the bulk of interactions occurs within modules rather than across module boundaries.

The idea of a collection of interdependent elements is common to many traditional definitions of the term system. In a complex system, it is dynamic interactionism rather than interdependence that matters (Lane, 2011). The elements constituting the complex systems interact so as to produce an outcome. This, in turn, is usually referred to as an "emergent property" of the system under scrutiny. The principle of emergence can be more sharply defined as the arising of novel and coherent patterns during a process of self-organization (Corning, 2002). Saviotti (2011) synthesizes the common characteristics of emergence as follows: radical novelty, coherence, global level, dynamics and ostensive. An emergent property is likely to be something which has not been observed in the past in the system and which shows a high degree of persistent integration in its components. By definition, it stems from the self-organization achieved through dynamic interactions of complex structures, therefore it is not reducible to the single components and can be understood only as a whole. Finally an emergent property is the product of a dynamic process which evolves over time. The combination of these properties makes it impossible to predict the attributes of emergent properties.

The better understanding of such dynamics requires the adoption of "organization thinking". Within the domain of complexity theory, by organization one means particular kinds of interacting entities which can be characterized by structure, function and process (Lane *et al.*, 2009; Lane, 2011). The concept of structure occurs quite frequently when talking about complexity. The structure of interactions among components is of key importance in shaping the emergence of new patterns. Following Simon (1962, 1973), complex systems can therefore be thought of as a (self-)organized set of interacting elements. The components in turn, understood as "organizations", are characterized themselves by a structure which consists of interacting elements, which in turn have their own structure. As Arthur (2009) and Lane (2011) emphasize, what was missing in Simon was this principle of recursivity. Thus complex systems are not only characterized by nested hierarchies, but each layer is organized according to the same principle of interacting components that interact in complex ways.

A crucial dimension to think about complex dynamics in social systems is the agent/artefact space. The contributions by Arthur (2009) focus primarily on the artefact side, and propose a concept of technology as the outcome of a combinatorial process, according to which each technology is characterized by one or a few key principles interacting with many other complementary technologies in defining the functionalities. As such technologies are characterized by an operational structure which organizes the way the components are combined. Such structures feature the components at each level of the nested hierarchy. New functionalities arise in the context of a process of exaptive bootstrapping. In other words, new artefacts, say a new product or new technologies, are designed to achieve some particular functionality. However, besides the expected functionality, the potentials of an artefact can be fully grasped only as a function of use. In the course of utilization of artefacts new patterns of interactions emerge around them, which leads to the emergence of new functionalities that can represent the main functionalities driving the design of further new artefacts. This in turn generates again patterns of utilizations likely conducive to the discovery of new functionalities, engendering a self-propelling process. Once new artefacts are introduced, they enter "in competition" with those already existing in the system, and therefore, since identity is relational,[2] the attribution of a new identity to the new artefact involves also the renegotiation of identity for previously existing ones. These new attributions emerge out of interactions that are called "generative relationships" (Lane, 2011).

In the agent/artefact space artefacts emerge as an outcome of the interactions among agents, who identify the functionalities to be implemented in the novel designs, and engage therefore in a combinatorial process. An important feature of complex systems is that the modification in one part of the structure is likely to be reflected on the other interconnected elements, as well as on the other structures linked through higher-level structural organizations. The implementation of a new functionality may require the modification of two or more attributions, what is called epistatic relationships. Moreover, one property of a system component is pleiotropy, i.e. the number of components that are likely to be changed as an effect of intentional changes on its role in the structure (this applies both to change of configuration and to substitution). The higher the pleiotropy of a particular component, the more risky is any action on it. Given a functioning design, the improvement of high-pleiotropy component may engender such a chain of adaptations in the structure that the net effect on the new design is a worsening of the whole performance.

Epistatic relationships and pleiotropy are features not only of the components of artefacts structure, but also of the "organizations" which represent the interacting entities in the agents' domain. People in a team, or firms in a joint research project, may be thought of as interacting elements of a complex system, which interact for the purpose of achieving a particular target which will come about as an emergent property. The metaphor of the network has been much used in the last decade to describe the structure of interactions of agents in complex socio-economic systems.[3] A network is indeed composed by nodes, i.e. the elements of a complex systems, and by edges, i.e. the actual interactive relationships among nodes. The usefulness of the network (and also a methodological tool, as we will see later on) lies in the fact that some of its properties depend exclusively on its geometry, or the architecture of its structure, irrespective of the characteristics of the nodes. Obviously, it provides also the means to describe the single nodes, although in a relational way. A particular class of networks is largely used in the context of complex system dynamics, i.e. "scale-free" networks (Barabasi and Albert, 1999; Barabasi *et al.*, 2002), which is characterized by a highly asymmetric distribution of links about nodes. Letting the degree of the number of links insisting on a node, scale-free networks are such that the degree distribution follows a power law. The existence of such distribution has been explained by adopting basically two mechanisms, i.e. fitness models and preferential attachment. The former mechanism (Bianconi and Barabasi, 2001) is based on the idea that new nodes entering a network choose the nodes with which establish a link on the basis of fitness values. In this direction, following a kind of "supremacy of the fittest" principle, nodes showing higher levels of fitness degree are likely to attract a higher number of links. The mechanisms of preferential attachment was formerly introduced by de Solla Price (1965), who talked about cumulative advantage in his scientometrics work. Barabasi and Albert (1999) have recently developed a model to model the growth of the World Wide Web based on this mechanism. The preferential attachment refers to a class of stochastic process in which some quantity is distributed among a number of individuals according to how much they already have. In other words, the flow is a function of the cumulated stock, such that the individuals showing the highest values of cumulated stock are interested by highest values of flow. According to this, the few nodes in the network showing high degree centrality are likely to increase their degree much more than the peripheral nodes.

The concept of network, and especially of scale-free networks, is particularly useful for social sciences, along with the properties allowing for their description. However, the discussion conducted so far proposes the existence of isomorphism between the structures in the agents and in the artefacts' domain. The network metaphor can be used therefore to describe also the structure of "organizations" in the artefact side, as well as the dynamics of their interactions. The possibility of using the same representation to represent the interactive dynamics of agents as well as the interactive dynamics of the emergent properties that they generate, along with the notions of hierarchy and recursivity, provides a powerful toolkit to understand changes in economic structure and changes in knowledge in an integrated framework. The following section will articulate a complex system perspective on structural change and knowledge structure by relying on a peculiar acceptance of the term "space".

4.5 Economic and knowledge structures: interacting subsystems in a nested hierarchy

The main path guiding the writing of this volume started by showing the empirical relevance of structural change in economics, in order to arrive at proposing the dynamic concept of knowledge structure according to which knowledge is an element in a wider system featured by complex dynamics, and characterized in turn by a structure with its own architecture. In this perspective knowledge has a structure in the same way as the economy. Both structures are related through a thick network of complex and dynamic interactions across different layers of a nested hierarchy.

An important brick in such building is, in our opinion, a contribution much neglected by scholars interested in grafting complex system dynamics onto economics, i.e. François Perroux' (1950) elaboration on the concept of economic space. Perroux maintains that "we may distinguish in our discipline as many economic spaces as there are constituent structures of abstract relations which define each object of economic science" (1950, p. 91). What is interesting is the definition of space on the basis of structure relations. Moreover, the idea of a multiplicity of spaces points to a set of interconnected structures. The author distinguishes between geonomic space and economic spaces. The former, also called "banal space", is defined by geonomic relations between points, lines and volumes. As an example, the characterization of firms on the basis of their geographical coordinates is based on their localization in a geonomic space. The same applies for two points in a Cartesian coordinate system. Economic spaces are instead "defined by the *economic relations* which exist between elements. These economic spaces conveniently reduce to three. (1) economic space *as defined by a plan*; (2) economic space *as a field of forces*; (3) economic space *as a homogenous aggregate*" (Perroux, 1950, p. 94). The first dimension refers to the set of relations among the units in the economic space. The second refers to the existence of centres, or poles, from which centrifugal forces emanate and to which centripetal forces are attracted. The concept of attractor, very much related to phenomena of persistence and therefore dynamic preferential attachment, can be included in this perspective. Finally, "the relations of homogeneity which define economic space ... are relative to the units and to their structure, or relative to the relations between these units" (Perroux, 1950, p. 96).

Perroux' elaboration of the concept of economic spaces allows for the integration of socio-economic phenomena into a single framework susceptible to being modelled as a complex system of interacting elements. Articulating socio-economic life in one single scheme is a pretty hard task which is beyond the scope of this volume. An illustrative example is provided in Figure 4.1, where we have partitioned the agents' and the artefacts' dimensions, and depicted the relationships between the most relevant subsystems as well as a sketch of the structure of the interactive elements they consist of. The artefact space is instead populated by objects the creation of which stems from an emergence process at the agents' levels. For the sake of clarity, we have omitted the exemplification of the complex structure that features each of these classes of artefact or lower-level subsystems in the agents' space. Firms, for example, can be described as networks in which nodes are represented by tasks/agents and the links by the transfers among the nodes (Baldwin, 2007).

There are some emergent properties that are not properly classifiable as artefacts, like knowledge or human capital, which can be thought of as the outcome of complex system dynamics of the agents acting in the education system as well as in families. There is also a sensible degree of overlap among the subsystems: firms, for example, are part of both the innovation and the productive system. Human capital is a structured component informing the productive system, the innovation system, the institutional system and the education system itself. By looking at this quite simplified diagram, it clearly emerges how each subsystem can be seen as a component of a higher-order system. After all, Perroux himself

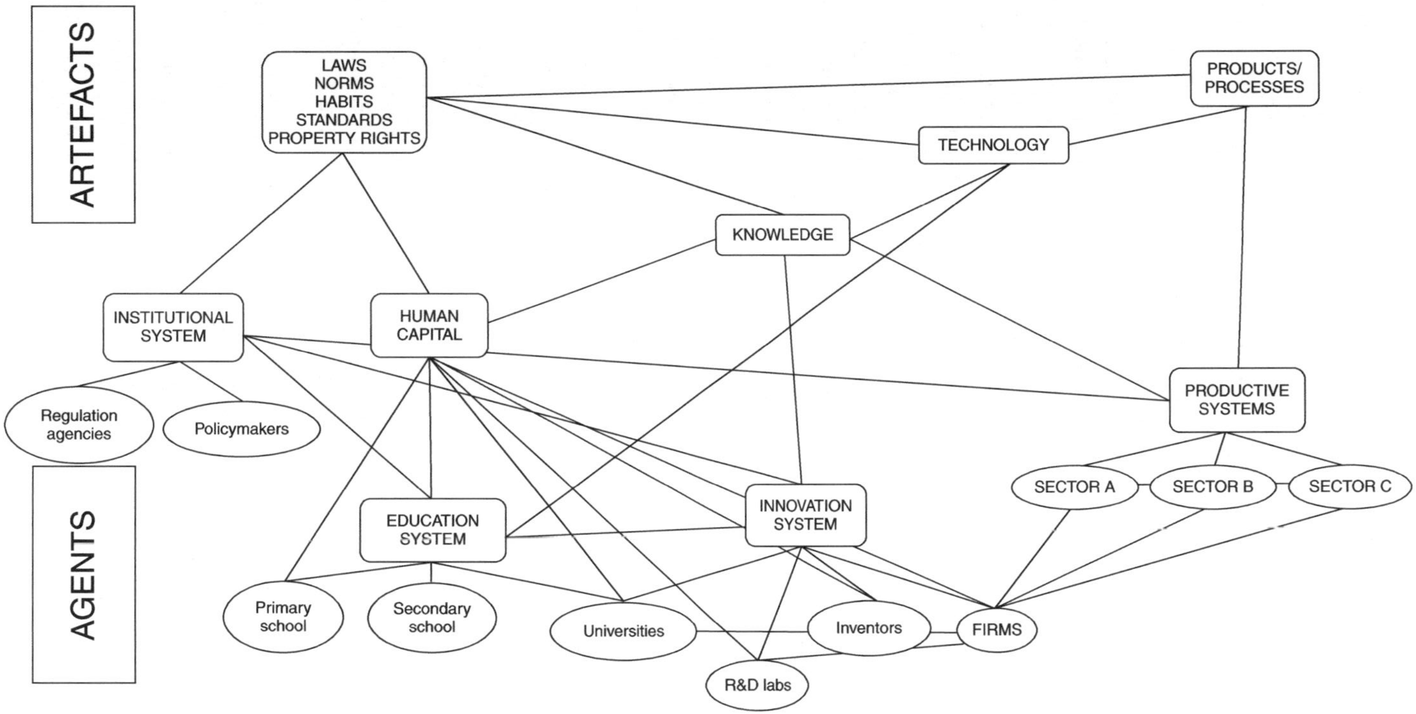

Figure 4.1 Simplified scheme of dynamic interactions in complex socio-economic systems.

emphasized the mutual dependence of different economic spaces. The organization of socio-economic systems seems therefore not to escape Gödel's theorem of incompleteness, according to which no system can be found able to be completely self-explaining.

The topology, or the structure, of relations occurring in such abstract spaces dominated by both between- and within-system complex interactions exhibits an architecture which shapes both the pattern of linkages across the components and the quality of the components themselves. The architecture is therefore a key concept for the analysis of complex dynamics. Henderson and Clark (1990) introduced the concept of architectural change in the context of product design complexity. The isomorphism which we maintain to characterize both artefacts' and agents' structures allows for extending the idea of architectural change beyond the scope of product technologies. The architecture of systems of interacting innovating agents is important in that it influences the likelihood to capitalize knowledge externalities and generate new technologies. Cowan and Jonard (2003) shows that the way the structure of interactions is designed has a strong influence on the system performance. Moreover, interactions across components are not equally productive. There are some components that are better suited to interact with other specific components. Network theorists have labelled this property as "homophily" of nodes (Skvoretz, 1991; Powell *et al.*, 2005). According to this principle, elements in a network are more likely to interact with other elements that are similar. This principle has been proposed as an explanation o the patterns of development of nations by Hidalgo *et al.* (2007) and Hidalgo (2009), who proposed the concept of product space conceived as a network in which nodes are product classes and links are the interactions among them. They show that the development pattern of nations is such that they move in the product space by developing goods that are close to what they already produce. The same principle underlies the idea that social proximity shapes the interactions of collaborative networks for innovation to a larger extent than geographical proximity (Ponds *et al.*, 2010).

The architecture of the structure of interactions in complex systems is therefore characterized by the patterns of linkages among components, as well as by the features of the components themselves. This supports the idea that the complexity approach is able to synthesize individualism and holism. A much neglected aspect of architecture is its number of components. In Henderson and Clark (1990) architectural change only concerns the changes in the patterns of relationships among components. One property of scale-free networks is the growth of the network itself, which occurs by the entry of new elements in the system. Architectural change can also happen in view of an addition of new components to the structure of relations. Altenberg's models of constructional selection, which extend Kauffman's N-K fitness landscape models, represent an interesting exception to the substantial neglect of endogenous change of architecture in the analysis of complex system dynamics.

To synthesize, the architecture of a complex system may well change over time, and so may the structure of epistatic *relationships*. This may occur either due to a change in the relative weight of some elements in the system, these elements switching from a non-influential to an influential position, or by means of introduction of new elements within the system. This is in turn likely to alter the existing structure of relationships. Within this context, the *pleiotropy* represents the number of elements in the system that are affected by the appearance of new elements. It is clear that the higher the pleiotropy, the greater the change in the architecture of the system that the inclusion of new elements may engender.

The viewpoint of endogenous complexity makes the analysis of knowledge dynamics particularly appealing and challenging. In view of the discussion conducted so far, *knowledge* can indeed be represented as an *emergent property stemming from multi-layered complex dynamics*. Knowledge is indeed the result of a collective effort of individuals who interact with one another, sharing their bits of knowledge by means of intentional acts of communication (Antonelli, 2008; Saviotti, 2007). In other words, the adoption of an endogenous complexity made possible by an augmented recombination approach allows for the combination of the view on technology as an artefact with the view of technology as an act, i.e. as the product of collective actions involving agents with converging

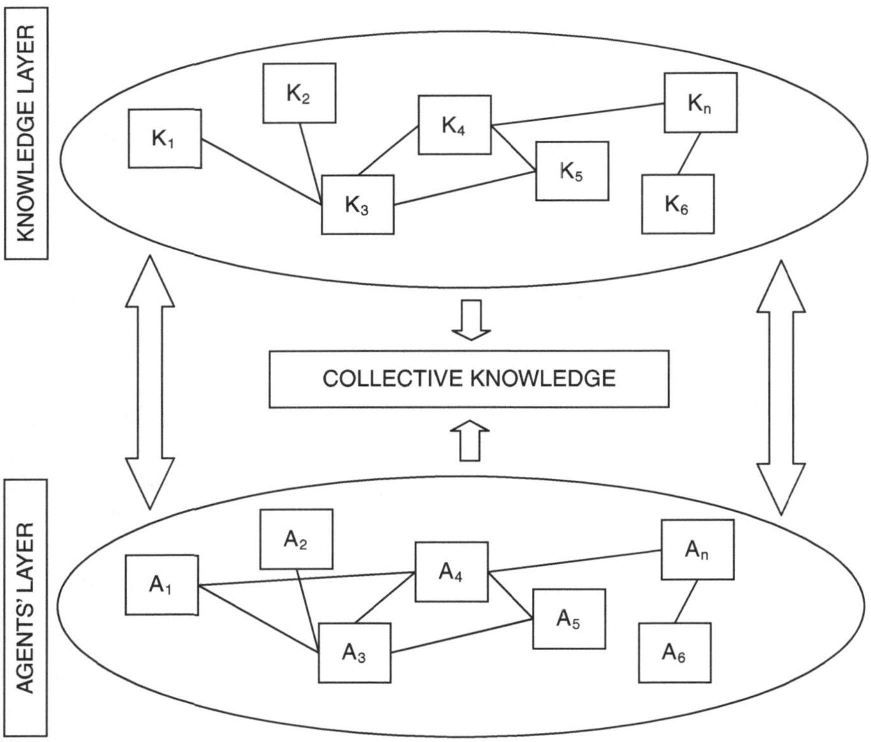

Figure 4.2 Zoom in the dynamic interactions between the knowledge and the innovation systems.

incentives and aligned interests (Figure 4.2 provide a zoom in the dynamic interactions between these two systems) (Arthur, 2009; Lane *et al.*, 2009).

The structure of the network of relationships among innovating agents represents therefore a crucial factor able to shape the ultimate outcome of knowledge production processes. Constructional selection matters, in that new institutions entering the network need first of all to choose which incumbents they want to be linked with. The concept of preferential attachment applies to this situation. In a wide number of contexts, the new nodes in a network generally end up linked with those "old" nodes already characterized by a large number of connections (Barabasi and Albert, 1999). As a consequence, the entrance of new actors in the network is likely to reshape the relative weight of nodes, and hence modify the structure and the balance of relationships.

Knowledge so produced stems from the combination of bits of knowledge dispersed among innovating agents. Creativity refers to the ability of agents to combine these small bits of knowledge so as to produce an original piece of technological knowledge. This in turn may be thought of as a structure of bits of knowledge linked with one another. The knowledge base itself, at whatever level, can be therefore imagined as a network in which the nodes are the small bits of knowledge and the links represent their actual combination in specific tokens. Knowledge in this sense turns out to be an emergent property of complex dynamics featuring the interdependent elements of the system, i.e. the bits of knowledge.

This is a quite unexplored consequence of the structural character of knowledge production, which provides further richness to its dynamics. Since such complex system may be represented as a network, the knowledge base is characterized by a structure with its own architecture.[4] This in turn may evolve over time, as an effect of the introduction of new small bits of knowledge and the consequent change in the relative weight of the nodes within the network, as well as due to the change in the patterns of linkages among bits of knowledge. Indeed, like in the networks of innovators, new nodes will be attached to some existing nodes, the centrality of which will be altered. Learning dynamics and absorptive capacity represent a channel through which the topology of knowledge structure affects search behaviour at the level of agents' networks. Indeed, agents move across the technology landscape in regions that are quite close to the area of their actual competences (principle of homophily). Technological change is localized as an effect of the interactions between the complex dynamics at the knowledge and the agents' level. However, the topology of knowledge structure is in turn shaped by the choices made by innovating agents as to which bits of knowledge combine. A self-sustained process is likely to emerge, according to which the knowledge creation process tends more and more towards a local attractor in which they are locked in (Colombelli and von Tunzelmann, 2011).

This dynamic indeed makes preferential attachment work also at the knowledge level. Agents' search behaviour is indeed constrained by the topology of the knowledge structure. In this direction, those small bits of knowledge which have grown in importance are likely to exert a much stronger influence. This process is rooted in historical time, according to which the gradual sorting out of knowledge bits which have proved not to be so fertile leaves the floor to fewer and more fertile bits. New bits of knowledge entering the knowledge base later on are likely to be linked to these few pillars.

Preferential attachment introduces a great deal of path dependence in system dynamics of technological knowledge. It amounts to articulating the concept of persistence beyond the rate of introduction of innovations, so as to apply it to the centrality of the specific smaller bits of knowledge which make the structure of the knowledge base.

Still, while this self-enforcing process is likely to trap the search process within a bounded area, the dynamics of technological communication at the agents' level as well as the capabilities to cope with search in areas that are far from the competences of innovating agents are likely to introduce discontinuities in the evolutionary pattern. This amounts to introducing a wide variety of new bits of knowledge which are loosely related with those already existing in the knowledge base, so as to give rise to radically new combinations. The process of evolution, fed by learning dynamics and cumulativeness, leads to the gradual selection of the best combinations (principle of fitness), which grow in centrality and hence begin to constrain agents' search behaviour. Knowledge sharing and technological communication ensure therefore the emergence of new variety, which is more likely to occur in transition phases. At this stage a wide range of alternatives are viable, and multiple local attractors are likely to emerge from mutual influences between complex dynamics at the knowledge and the agents' layers.

Clearly, the patterns of change in the architecture are likely to bear important systemic effects. First of all, the impact of node substitution depends on the pleiotropy level. Changes affecting a high-pleiotropy node by definition will engender changes and adaptation in a large part of the system. The change of a high-pleiotropy node in the structure of knowledge is likely to generate a discontinuity in the knowledge base. Interestingly enough, a node, a small unit of knowledge, can be co-responsible for its own substitution. Knowledge bits are indeed combined to create new knowledge. This new knowledge can also cause the elimination or the improvement of one of the bits used to generate it. Generative relationships matter in that each module in the knowledge structure is likely to interact to generate new modules.

The introduction of a discontinuity in the knowledge base can be reflected in the connectivity or in the density of the network, and also have important consequences on the interaction dynamics of innovating agents. The introduction of new agents can be necessary in order to command the effects of the discontinuity. Obviously, these new agents will need to choose which of the existing nodes in the network they want to be attached to. Preferential attachment and fitness values are clearly important in this respect. New entrants will be more prone to be connected to the incumbents showing the better performances, at least in the early stages of the paradigm change. The dynamics of discontinuities and technological alliances in the biotechnology sector are particularly exemplary of such phenomena (Krafft *et al.*, 2011). Search strategies in these contexts are initially oriented towards the exploration of the technological opportunities provided by the new combinatorial potentials, but this does not necessarily imply that innovating agents proceed in a scattered and random way in the knowledge landscape. Agents with high fitness values are better able to narrow the exploration patterns in paths that are not too distant from the core of the cumulated technological competences (Colombelli *et al.*, 2011).

Of the systemic effects generated by changes in the architecture of knowledge structure, of particular relevance are the interaction dynamics with the economic structure. This can indeed be viewed as an element of the higher-order system which is mutually interdependent with the other subsystems. Even though, according to the principle of near-decomposability, interactions within subsystems should be more frequent than those between them, Figure 4.1 shows clearly that it is quite difficult to identify sharp boundaries across subsystems, as a large number of overlapping regions can be devised. The dense chain feedbacks between economic and knowledge structure (and vice versa) after all are not surprising for industrial economists. Indeed, as emphasized in Chapter 2, since the early contributions of Kuznets (1930), Burns (1934) and Schumpeter (1939), the emergence and evolution of new industries has been depicted as phenomena strictly linked to the dynamics of technological change. In the footsteps of these seminal works, a wide body of literature has emphasized the interplay between technology and industry evolution (Agarwal and Tripsas, 2008). The concept of lifecycle plays a key role in this respect. On the one hand, industry lifecycles are described in terms of firms' patterns of entry and exit, as well as of changing industrial structure and size distribution over time. On the other hand, the technological lifecycle is described in terms of different patterns of firms' search behaviour, and the introduction of radical rather than incremental technological change (Dosi, 1982; Malerba and Orsenigo, 1995; Klepper, 1997).

The technological lifecycle literature posits that most of the development of many new industries is shaped by underlying evolutionary changes in technological knowledge. In this framework new industries emerge out of the introduction of technological discontinuities. The emergence of a discontinuity which occurs at the beginning of a new technological paradigm is likely to be accompanied by phenomena like the replacement of industrial firms linked to the old paradigm by new ones or by considerable changes in the personnel and in the internal organization of incumbent firms. Schumpeter (1939) had already noticed how the producers of cars or trains would generally not be the same firms which had produced mail coaches. In more modern terms such phenomena can be explained by the competence-disrupting effect of the discontinuity inherent in radical innovations (Tushman and Anderson, 1986).

The extent of discontinuity and its impact on industrial organization is highest in the early phases of a new technological paradigm and to become gradually less important as the paradigm matures. This lifecycle is likely to be accompanied by a number of changes such as: (a) transition from competence-disrupting to competence-enhancing technological change; (b) movement from exploration to exploitation (March, 1991); (c) movement from random search to organized search (Krafft *et al.*, 2009).

4.6 Conclusions

This chapter has articulated an integrated approach to knowledge structure and to its relationships with the economic structure, by adopting a complex system perspective. We have proposed a path moving from the appreciation of the interactive dynamics of innovation, so as to emphasize the combinatorial process underlying the generation of new technological knowledge. The revival of the concept of "structure" has proved to be fertile for our purposes, and in particular the integration of the genetic structuralism in the development of the framework. The historical approach to the generation and modification of structures allows us to grasp their evolutionary dynamics. Complex system dynamics enter the discourse as a heuristic able to overcome the excessive holism of structuralism by paying attention also to the features of the elements composing the structure. We have discussed some important concepts for complexity theory, like emergence, epistatic relationships, pleiotropy, scale-free networks, exaptive bootstrap and generative relationships, and showed how these can be useful to derive an approach to the analysis of the socio-economic system as a nested hierarchy characterized by recursive structures. The interplay between knowledge structure, economic structure and innovation systems can therefore be better grasped by adopting such an integrated view based on dynamic interactions generating unpredictable outcomes having sound systemic effects.

With this framework in mind, we can now move the concrete analysis of economic reality, by deriving some implications in terms of methodology of analysis, and then trying to integrate such view into treatable empirical frameworks.

Notes

1 Nicolas Bourbaki used to be a collective pseudonym (like the Wu Ming) of a group of French mathematicians working, since 1935, on a definitive survey of Mathematics (see Corry, 1992).
2 Once again a key concept for structuralists.
3 This is hardly surprising in the age of ICTs. According to McLuhan (1964), indeed, societies are always informed more by the means used to convey information than by the content of the information itself. De Kerckhove (1997) argues that the fundamental stages in humans' "cognitive" development correspond to the ways in which communication techniques and technologies have shaped not only interpretations of the human mind and brain function, but also views of society and the world over the ages. The paradigm shift towards network-based theories is therefore to be associated also with the pervasive diffusion of ICTs into human life both for individual and social purposes.
4 Although in a more orthodox framework, Olsson's (2000) and Olsson and Frey's (2002) concept of knowledge as a (convex) set in the idea space provides interesting insights.

Part III

Applications

5 The implementation of knowledge structure

Methodological implications

5.1 Introduction

The possibility of representing knowledge as a network provides an adequate conceptual foundation for the study of processes of knowledge generation and utilization in firms and industries. To identify all the variables and the connections present in the knowledge base of a firm at the lowest possible level of aggregation would be a prohibitively expensive task. An approximate version can then consist of identifying relatively "small" units of knowledge and their connections. We identify these "small" units within the traces of knowledge which have been used so far, such as patents and publications.

At whichever level of analysis, the knowledge base (KB) can be defined as the collective knowledge that agents can use to achieve their productive objectives. The collective character comes from the interactions between individuals, research units and departments of the same firm or research organization. Such interactions are specific to each organization and can be expected to lead to a different knowledge time path even in a case in which the initial competences of all the persons employed were the same. When we want to study the knowledge base of an industrial sector or of a field of science, such collective character of course includes inter-organizational interactions.

The KB can be mapped by identifying the units of knowledge composing it and by their connections or links. These units can be either technological classes or themes. Connections are determined by the joint utilization of the units in particular texts, be they patents, papers or something else. For example, if we use technological classes, the connections are given by the co-occurrence of different classes in the patents used, and the frequency of co-occurrence can be interpreted as a measure of the strength of the link. In this way we can construct visual maps of the KB of a firm and follow the evolution of such KB over time. These maps of the KB can be considered a representation of the brain of the firm.

In order for these maps not to be purely descriptive devices we need to identify some general properties of the KB which can be measured and used both in empirical studies and in modelling, by exploiting both the network structure of knowledge and the statistical potentialities provided by the matrices of technological co-occurrence.

5.2 The use of co-occurrence matrices: coherence, cognitive and variety

The three properties of the KB which we will use in our analysis are its variety, related or unrelated, its coherence and its cognitive distance.

The *variety* of a KB measures the extent of its diversification, with related variety measuring it at a lower level of aggregation and unrelated variety at a higher level of aggregation (Frenken *et al.*, 2007). Technological variety can be measured by using the information entropy index. It was introduced by Shannon (1948) to measure the information content of messages, and can be used as a distribution function in a number of circumstances (Theil, 1967; Frenken and Nuvolari, 2004). The use of information entropy to measure variety is based on the rise in the information content of systems as the number of their distinguishable components increases: a system with a large number of distinguishable components requires more information to be described than a system with a smaller number of distinguishable components.

The information entropy index has interesting features, like its decomposability into a between-group and within-group component, and the extension to multidimensional cases. According to the latter, one may calculate the variety of the actual combinations of technological classes in a given context (say a firm or a sector). The former property allows for the operationalization of the distinction between related and unrelated variety. One could say that related variety (within-group entropy) measures diversification at a local level, or within a technological class, while unrelated variety (between-group entropy) measures diversification at a more global level in a knowledge space. The important implication of this distinction is that while a growth in unrelated variety implies a rise in cognitive distance, a growth in related variety is compatible with a more incremental development and even a fall in cognitive distance.

The *coherence* of a KB measures the extent to which different types of knowledge can be combined. This is of fundamental importance since the types of knowledge required by firms to create new products or services are not necessarily found within a discipline, but need to be combined to produce the desired output. The ability of firms to combine these different types of knowledge is not constant but can be expected to vary systematically during particular phases of the evolution of knowledge. For example, we can expect the ability of firms to combine different types of knowledge to fall as a completely new type of knowledge emerges at a discontinuity and to rise again as the new type of knowledge starts maturing. The coherence of the knowledge base can be calculated by modifying a procedure developed by Teece *et al.* (1994) to measure the coherence in the output of a firm. The basic principle underlying the calculations is that the higher the frequency that different technologies are used together by a firm, the more coherent is its knowledge base. The calculation proceeds by first calculating the frequency of co-occurrence of each pair of technologies in the KB and then by averaging them over the whole firm, or sector in the present case (see Nesta and Saviotti, 2005, 2006; Krafft *et al.*, 2009).

Cognitive distance measures the extent of discontinuity involved in the emergence of a new type of knowledge. It is the inverse of an index of similarity. This measure is of fundamental importance to be able to distinguish the effect of the emergence of a discontinuity from that of the subsequent period of normal or incremental development. There are many ways to calculate cognitive distances but we used the complement of the index of similarity proposed by Jaffe (1989).

From a technical viewpoint, such variables will be implemented throughout the analyses in this book as follows.

5.2.1 An overview on calculations

5.2.1.1 Variety

Let us start with the variety indicator, which we decided to measure using the information entropy index. Entropy measures the degree of disorder or randomness of the system, so that systems characterized by high entropy will also be characterized by a high degree of uncertainty (Saviotti, 1988). Differently from common measures of variety and concentration, the information entropy has some interesting properties (Frenken and Nuvolari, 2004). An important feature of the entropy measure is its multidimensional extension. Consider a pair of events (X_l, Y_j), and the probability of co-occurrence of both of them p_{lj}. A two dimensional total variety (*TV*) measure can be expressed as follows:

$$TV \equiv H(X,Y) = \sum_{l}\sum_{j} p_{lj} \log_2\left(\frac{1}{p_{lj}}\right) \tag{5.1}$$

If one considers p_{lj} to be the probability that two technological classes l and j co-occur within the same patent, then the measure of multidimensional entropy focuses on the variety of co-occurrences of technological classes within regional patents applications.

Moreover, the total index can be decomposed in a "within" and a "between" part any time the events to be investigated can be aggregated into a smaller numbers of subsets. Within-entropy measures the average degree of disorder or variety within the subsets, while between-entropy focuses on the subsets measuring the variety across them. Frenken *et al.* (2007) refer to between- and within-group entropy respectively as unrelated and related variety.

It can be easily shown that the decomposition theorem holds also for the multidimensional case. Hence if one allows $l \in S_g$ and $j \in S_z$ ($g = 1,\ldots, G$; $z = 1,\ldots, Z$), we can rewrite $H(X, Y)$ as follows:

$$TV = H_Q + \sum_{g=1}^{G}\sum_{z=1}^{Z} P_{gz} H_{gz} \tag{5.2}$$

Where the first term of the right-hand side is the between-entropy and the second term is the (weighted) within-entropy. In particular:

$$UTV \equiv H_Q = \sum_{g=1}^{G} \sum_{z=1}^{Z} P_{gz} \log_2 \frac{1}{P_{gz}} \tag{5.3}$$

$$RTV \equiv \sum_{g=1}^{G} \sum_{z=1}^{Z} P_{gz} H_{gz}$$

$$P_{gz} = \sum_{i \in S_g} \sum_{j \in S_z} p_{lj} \tag{5.4}$$

$$H_{gz} = \sum_{i \in S_g} \sum_{j \in S_z} \frac{p_{lj}}{P_{gz}} \log_2 \left(\frac{1}{p_{lj} / P_{gz}} \right)$$

We can therefore refer to between- and within-entropy respectively as *unrelated technological variety* (*UTV*) and *related technological variety* (*RTV*), while total information entropy is referred to as *general technological variety*.

5.2.1.2 Knowledge similarity and dissimilarity (cognitive distance)

We need a measure of cognitive distance (Nooteboom, 2000) able to express the dissimilarities among different types of knowledge. A useful index of distance can be derived from the measure of *technological proximity*. Originally proposed by Jaffe (1986, 1989), who investigated the proximity of firms' technological portfolios. Subsequently Breschi *et al.* (2003) adapted the index in order to measure the proximity, or relatedness, between two technologies. The idea is that each firm is characterized by a vector V of the k technologies that occur in its patents. Knowledge similarity can first be calculated for a pair of technologies l and j as the angular separation or uncentred correlation of the vectors V_{lk} and V_{jk}. The similarity of technologies l and j can then be defined as follows:

$$S_{lj} = \frac{\sum_{k=1}^{n} V_{lk} V_{jk}}{\sqrt{\sum_{k=1}^{n} V_{lk}^2} \sqrt{\sum_{k=1}^{n} V_{lk}^2}} \tag{5.5}$$

The idea underlying the calculation of this index is that two technologies j and l are similar to the extent that they co-occur with a third technology k. The cognitive distance between j and l is the complement of their index of the similarity:

$$d_{lj} = 1 - S_{lj} \tag{5.6}$$

Once the index is calculated for all possible pairs, it needs to be aggregated at the industry level to obtain a synthetic index of technological distance. This can be done in two steps. First of all one can compute the weighted average distance of technology l, i.e. the average distance of l from all other technologies.

$$WAD_{lt} = \frac{\sum_{j \neq l} d_{lj} P_{jit}}{\sum_{j \neq l} P_{jit}} \tag{5.7}$$

Where P_j is the number of patents in which the technology j is observed. Now the average cognitive distance at time t is obtained as follows:

$$CD_t = \sum_l WAD_{lit} \times \frac{P_{lit}}{\sum_l P_{lit}} \tag{5.8}$$

5.2.1.3 Knowledge coherence

Cognitive distance measures the degree of dissimilarity among technologies. We expect it to provide us with an indication of the difficulty, or cost, a firm has to face to learn a new type of knowledge. Typically a firm needs to combine, or integrate, many different pieces of knowledge to produce a marketable output. Thus, in order to be competitive a firm not only needs to learn new "external" knowledge but it needs to learn to combine it with other, new and old, pieces of knowledge. We can say that a knowledge base in which different pieces of knowledge are well combined, or integrated, is a coherent knowledge base. The technologies contained in the knowledge base are by definition complementary in that they are jointly required to obtain a given outcome. For this reason, we turn to calculating the coherence of the knowledge base, defined as the average relatedness of any technology randomly chosen within the sector with respect to any other technology (Nesta and Saviotti, 2005, 2006; Nesta, 2008).

To yield the knowledge coherence index, a number of steps are required. In what follows we will describe how to obtain the index at whatever level of analysis i. First of all, one should calculate the weighted average relatedness WAR_j of technology j with respect to all other technologies present within the sector. Such a measure builds upon the measure of technological relatedness τ_{jm} (see below). Following Teece *et al.* (1994), WAR_j is defined as the degree to which technology j is related to all other technologies $j \neq m$ in the aggregate, weighted by patent count P_{mt}:

$$WAR_{jit} = \frac{\sum_{m \neq j} \tau_{jm} P_{mit}}{\sum_{m \neq j} P_{mit}} \tag{5.9}$$

Finally the coherence of knowledge base within the aggregate i (be it a firm, a sector or a region) is defined as weighted average of the WAR_{lt} measure:

$$R_{it} = \sum_{j \neq m} WAR_{jit} \times \frac{P_{jit}}{\sum_j P_{jit}} \tag{5.10}$$

It is worth stressing that such index implemented by analysing co-occurrences of technological classes within patent applications measures the degree to which the services rendered by the co-occurring technologies are complementary to one another. The relatedness measure τ_{jm} indicates indeed that the utilization of technology j implies that of technology m in order to perform specific functions that are not reducible to their independent use. This makes the coherence index appropriate for the purposes of this study.

In order to calculate the parameter τ, i.e. *technological relatedness*, we start by calculating the relatedness matrix (Nesta, 2008). The technological universe consists of k patent applications. Let $P_{jk}=1$ if the patent k is assigned the technology j [$j=1,\ldots,$ n], and 0 otherwise. The total number of patents assigned to technology j is $O_j=\Sigma_k P_{jk}$. Similarly, the total number of patents assigned to technology m is $O_m=\Sigma_k P_{mk}$. Since two technologies may occur within the same patent, $O_j \cap O_m \neq \varnothing$, and thus the observed the number of observed co-occurrences of technologies j and m is $J_{jm}=\Sigma_k P_{jk} P_{mk}$. Applying this relationship to all possible pairs, we yield a square matrix $\Omega(\mathrm{n} \times \mathrm{n})$ whose generic cell is the observed number of co-occurrences:

$$\Omega = \begin{bmatrix} J_{11} & & J_{j1} & & J_{n1} \\ \vdots & \ddots & & & \\ J_{1m} & & J_{jm} & & J_{nm} \\ \vdots & & & \ddots & \\ J_{1n} & \cdots & J_{jn} & \cdots & J_{nn} \end{bmatrix} \tag{5.11}$$

We assume that the number x_{jm} of patents assigned to both technologies j and m is a hypergeometric random variable of mean and variance:

$$\mu_{jm} = E(X_{jm} = x) = \frac{O_j O_m}{K} \tag{5.11}$$

$$\sigma_{jm}^2 = \mu_{jm} \left(\frac{K - O_j}{K} \right) \left(\frac{K - O_m}{K - 1} \right) \tag{5.12}$$

If the observed number of co-occurrences J_{jm} is larger than the expected number of random co-occurrences μ_{jm}, then the two technologies are closely related: the fact the two technologies occur together in the number of patents x_{jm} is not casual. The measure of relatedness hence is given by the difference between the observed number and the expected number of co-occurrences, weighted by their standard deviation:

$$\tau_{jm} = \frac{J_{jm} - \mu_{jm}}{\sigma_{jm}} \tag{5.13}$$

It is worth noting that such relatedness measure has lower and upper bounds: $\tau_{jm} \in]-\infty; +\infty[$. Moreover, the index shows a distribution similar to a t-student, so that if $\tau_{jm} \in]-1.96; +1.96[$, one can safely accept the null hypothesis of non-relatedness of the two technologies *j* and *m*. The technological relatedness matrix Ω' may hence be thought of as a weighting scheme to evaluate the technological portfolio of regions.

The adoption of these variables marks an important step forward in the operational translation of knowledge creation processes. In particular, they allow for a better appreciation of the collective dimension of knowledge dynamics. Knowledge is indeed viewed as the outcome of a combinatorial activity in which intentional and unintentional exchange among innovating agents provides the access to external knowledge inputs (Fleming *et al.*, 2007). The network dynamics of innovating agents provide the basis for the emergence of new technological knowledge, which is in turn represented as an organic structure, characterized by elementary units and by the connections among them. The use of such variables implies therefore a mapping between technology as an act and technology as an artefact (Arthur, 2009; Lane *et al.*, 2009; Krafft and Quatraro, 2011). Co-occurrences matrices are very similar to design structure matrices (DSM) (Baldwin and Clark, 2000; Murmann and Frenken, 2006; Baldwin, 2007), in that they can be thought of as adjacency matrices in which we are interested not only in the link between the elements, but also by the frequency with which such links are observed.

In other words these measures capture the design complexity of knowledge structure, and allow for featuring the innovation behaviour of firms, as well as its evolution, in relation with the changing architecture of such structure (Henderson and Clark, 1990; Murmann and Frenken, 2006). In this perspective, an increase in knowledge coherence is likely to signal the adoption of an exploitation strategy, while a decrease is linked to exploration strategies. Increasing values of cognitive distance are instead related to random screening across the technology landscape, while decreasing cognitive distance is more likely to be linked to organized search behaviour. Knowledge variety is likely to increase in any case when new combinations are introduced in the system. However, the balance between related and unrelated variety should be such that the related one is likely to dominate during exploitation phases, while the unrelated one gains more weight in the exploration strategies (Krafft *et al.*, 2009).

5.3 Social network analysis[1]

The representation of the knowledge structure as a network clearly lends itself to the utilization of the toolkit provided by social network analysis. A *network* may be defined as a graph made of nodes that are tied to each other by one or more types of interdependency. Relationships among nodes are expressed by arcs, which in turn may be directed or undirected. Two nodes that are connected by a line are said to be *adjacent* to one another. Adjacency is therefore the graphical expression of the fact that two nodes are directly related or connected to one another. The points to which a particular point is adjacent are termed its *neighbourhood.*

Points may be directly connected by a line, or they may be indirectly connected through a sequence of lines. It may be thought of as a "walk" in which each point and each line are distinct. This is called *path*. The *length* of path is measured by the number of lines that constitute it. The *distance* between two points is the shortest path (the geodesic) that connects them.

One of the most widely used measures to describe a network is the *density*. It describes the general level of linkage among the points in a graph. The density of a network is therefore defined as the total number of actual lines, expressed as a proportion of the maximum possible number of lines:

$$\Delta = \frac{l}{n(n-1)/2} \tag{5.14}$$

A network is complete when all the nodes are adjacent, and the measure of density attempts to summarize the overall distribution of lines in order to assess how far the network is from completion. Density depends upon two other important parameters of the network, i.e. the inclusiveness and the sum of the degree of its points. *Inclusiveness* can be defined as the share of network nodes that are not isolated, i.e. the share of nodes that are connected to at least one other node. For example, in a network of 20 nodes with 5 isolated nodes the inclusiveness is 0.75. The more inclusive the graph, the more dense the network will be.

However, some nodes will be more connected than other ones. The *degree* of a node is an important measure of centrality that refers to the total number of other points in its neighbourhood. Formally one can represent the degree by the following equation:

$$D(v) = \sum_{s \in V \neq v} x_{vs} \tag{5.15}$$

This measure is obviously biased by the network size. Therefore it is useful to use a standardized measure, which consists in dividing the degree measure by its maximum value as follows:

$$ND(v) = \frac{D(v)}{n-1} \tag{5.16}$$

The higher the degree of the connected points in the network, the higher will be the density. For this reason the calculation of density needs to take into account both measures. It should compare the actual number of lines present in the graph with the total number of lines that the graph would show if it were complete.

While the density describes the network as a whole, the measures of *centrality* refer to the relevance of the nodes belonging to the network. A point is locally central if it has a large number of connections with other points in its immediate environments, i.e. other points in its neighbourhood. Global centrality refers instead to the prominence of the node with respect to the overall structure of the network. Measures of global and of local centrality have a different meaning.

Measures of global centrality are expressed in terms of the distance between various points. Two of these measures, i.e. closeness and betweenness, are particularly important. The simplest notion of *closeness* is that calculated from the "sum distance", the sum of geodesic distances to all other points in the graph (Sabidussi, 1966). After having calculated the matrix of distances between the nodes of the network, the sum distance is the row of column marginal value. A point with a low sum distance is close to a large number of other points, and so closeness can be seen as the reciprocal of the sum distance. Formally it can be expressed as follows:

$$C(v) = \frac{1}{\sum_{t \in V \neq v} d_G(v,t)} \tag{5.17}$$

where the denominator represents the sum of the geodesic distance of the vertex v to all other points.

The *betweenness* measures the extent to which a particular point lies "between" the other points in the graph: a point with a relatively low degree may play an important intermediary role and so be very central to the network (Freeman, 1979). The betweenness of a node measures how much it can play the part of a broker or gatekeeper in the network. Freeman's approach is built upon the concept of local dependence. A point is dependent on another if the paths which connect it to the other points pass through this point. Formally, let G be a graph with n vertices, then the betweenness is calculated as follows:

$$B(v) = \sum_{\substack{s \neq v \neq t \in V \\ s \neq t}} \frac{\sigma_{st}(v)}{\sigma_{st}} \tag{5.18}$$

where σ_{st} is the number of shortest geodesic paths from s to t, and $\sigma_{st}(v)$ is the number of shortest geodesic paths from s to t passing through a vertex v.

The centrality measures discussed above allow us to characterize each single network node. However, for the purposes of this chapter it is worth calculating the sector averages for all of the three indexes. In this direction, one must consider that each node corresponds to a technological class observed with a specific relative frequency, which must be taken into account when averaging out the centrality measures. We can then propose weighted average centrality measures as follows. Let $Z(v)$ be one of the three centrality measures referred to the generic node v, the weighted average centrality at time t is:

$$\overline{Z(v)} = Z(v) \times \frac{P_v}{\sum_v P_v} \tag{5.19}$$

where P_v is the number of patents in which the technology v is observed.

The huge potential of social network analysis lies in the possibility to map the yearly patterns of co-occurrences into network structures for the aggregate under scrutiny. This would allow to obtaining the dynamics of such indicators like density, connectivity or node centrality, so as to investigate the evolution of knowledge structure as well as the co-evolutionary patterns of other relevant structures in the hierarchy of nested subsystems.

5.4 Conclusions

The adoption of a structuralist approach to technological knowledge developed in Chapter 4 allows us to graft the analysis of knowledge into a complex-system dynamics framework. Knowledge can be accordingly viewed as a subsystem of a hierarchical organization, and its structure can be approximated by a network the nodes of which are the concepts that are combined in the process of knowledge generation, and the links are the actual combinations.

In this chapter we have attempted to investigate the methodological consequences of such approach, as far as the elaboration of consistent indicators is concerned. We have explored two distinct avenues that can be implemented by using the most popular proxies for scientific and technological outputs, i.e. patents and publications. We have showed how patent documents can provide useful information to implement both to derive indicators like coherence, cognitive and variety, which are based on the frequency by which couples of technological classes co-occur in the same patent, and to implement indicators which are part of the toolkit of social network analysis.

In the following chapters we will show how flexible these methodologies can be, in that they allow for investigating different phenomena, like evolutionary patterns of technological lifecycles or the differential effects of innovation behaviour on economic performances, at different levels of aggregation, be they sectoral, regional or national.

Note

1 This section builds on Scott (2000) and Wasserman and Faust (2007).

6 The internal structure of technological knowledge and productivity growth

Cross-country evidence from the ICT sector

6.1 Introduction

Since the seminal contributions by Schumpeter (1942), the analysis of the relationships between knowledge and innovation on the one hand, and economic growth on the other, has more and more attracted economic scholars. Empirical contributions estimating the relationship between knowledge and productivity have subsequently appeared thanks to path-breaking works by Zvi Griliches (1979). Most of them consisted of industry- or firm-level analyses,[1] while a much lower number of studies provided cross-country comparisons of the relationship between knowledge and productivity growth.[2] All these contributions shared an approach to technological knowledge as an unbundled stock. At present it is no longer sufficient to articulate the hypothesis that technological knowledge is a major factor in economic growth. More details and specifications are necessary to enquire into the specific forms of the relationship between the characteristics of the generation of technological knowledge and actual increases in rates of economic growth.

This chapter explores some key aspects of the generation of the technological knowledge that lies at the heart of the emergence of the new technological system based upon information and communication technologies (ICTs).[3] To this purpose, we combine the recombinant growth approach and the analysis of the role of variety in the economics of knowledge. We adopt Pier Paolo Saviotti's view of knowledge as a retrieval/interpretative and co-relational structure. This allows us to represent the knowledge base of the sector as a network whose nodes are constituted by technological classes, and to measure a number of properties of the knowledge base by means of co-occurrence matrices (Saviotti, 2004, 2007). We explore and identify a number of key characteristics of the recombinant generation of new technological knowledge and demonstrate their relevance for understanding the dynamics of economic growth.

We focus on the ICT sector knowledge base and its evolution through the 1980s and 1990s, and on its relationship with productivity growth in a sample of 14 representative OECD countries. The evolution of the ICT sector from its origins in the 1950s has been characterized by a process of continuous and rapid technological change, throughout which incremental innovation has been punctuated by major scientific breakthroughs (Bresnahan and Malerba, 1999). The development of ICTs can be represented as a typical Schumpeterian gale of innovation characterized by increasing convergence and the integration among a variety of localized innovations, generated within a wide range of industries and firms. Technological convergence has been driven by the introduction of a number of innovations such as Internet services, enhanced broadband fibre-optics, Asynchronous Digital Subscriber Lines (ADSL), digital television and a universal mobile telecommunications system, opening up the possibility of integrating a variety of content, services, technologies and applications (Fransman, 2002, 2007). As a result ICT, and the related technological knowledge, are analysed as a new technological system stemming from the recombination of a variety of knowledge modules that has fed an array of applications in many technologies favouring their rejuvenation (Quatraro, 2009b; Van den Ende and Dolfsma, 2005).

The evolution of the new technological system, marked by the increasing convergence of telecommunications and electronics during the 1980s, led to a reallocation of technological effort focused mainly, in the second half of the 1990s and the early 2000s, on the provision of content for the Internet and on wireless communication. Alongside this changing technological focus, the ICT ecosystem underwent a thorough reorganization of the international division of labour, with respect to the different layers in which it is articulated (Fransman, 2007; Krafft, 2004, 2009; Krafft and Salies, 2008).

The analysis of the generation and dissemination of ICTs in the last decades of the twentieth century therefore provides clear evidence on the working of recombinant knowledge: knowledge recombination is at the centre of the dynamics and is characterized by a clear sequence based upon a highly selective process of exploration (Corrocher *et al.*, 2007).

The contribution of this chapter to the existing literature is threefold. First, and most importantly, it provides a theoretical framework that implements and articulates the notion of recombinant knowledge for the analysis of the emergence of new technological systems. Second, it proposes a methodology based on the analysis of the co-occurrence of technological classes in one or more patents, to operationalize the empirical investigation of the recombination of different technologies. Third, it provides further support for the idea that, in order to assess the relationship between the generation of new knowledge and economic growth, the focus on knowledge capital stock and traditional indicators of its quality such as patent citations and litigations, is not sufficient to capture the qualitative changes that affect the internal structure of knowledge bases at firm level and at more aggregate levels of analysis.

The chapter is organized as follows. Section 6.2 articulates the research strategy, by introducing the knowledge-related measures that we maintain are better suited to the analysis of recombinant knowledge, and qualifies our working hypotheses. Section 6.3 describes the datasets used in this study and Section 6.4 presents the empirical evidence concerning the evolution of the knowledge-related measures across the sampled countries in the ICT field, while Section 6.5 shows the results of the econometric analysis. Section 6.6 provides a discussion of the main findings and offers some conclusions.

6.2 Research strategy

The argument elaborated so far leads us to maintain that new indicators of the quality of the knowledge portfolio of both firms and regions, industries or countries at more aggregate levels need to be elaborated, in order to gain a better assessment of the relationships between knowledge and productivity growth. Traditional indicators such as the knowledge capital stock or patent-based measures of knowledge quality are not sufficient. Work on assessing the quality of knowledge stocks based on such indicators as patent citations, infringements and litigation (Jaffe and Trajtenberg, 2002; Harhoff and Reitzig, 2004; Harhoff *et al.*, 2003) risks reflecting the effects of patent races and, hence, tends to dwell on the consequences of oligopolistic rivalry in product markets rather than the sheer quality of patents. Litigation and citations are much less relevant in emerging technological fields where oligopolistic rivalry has not become the dominant market form (Hall and Ziedonis, 2001, 2007).

On this basis we may therefore formulate a preliminary empirical specification to test the hypotheses spelled out in the previous section:

$$\ln\left(\frac{A_i(t)}{A_i(t-1)}\right) = a + b\ln A_{i,t-1} + \sum_n c_n \ln K_{n,i,t-1} + \sum_m d_m \ln Z_{m,i,t-1} + u \qquad (6.1)$$

According to Equation (6.1), A_{t-1} is the rate of multifactor productivity (MFP) growth of country i and it is a function of n characteristics of the knowledge base of the ICT sector and m control variables, with u being the error term (see Appendix 6.A for details on calculations of MFP growth rates). All the explanatory variables are lagged in order to reduce the risk of spurious correlations. Moreover, and as is usual in this type of empirical setting, we include in the structural equation the lagged level of productivity, ln $A_{i,t-1}$, in order to capture the possibility of mean reversion.

Our approach allows us to identify and measure a new qualification of technological knowledge. The exploration of the knowledge space enables us to qualify the distribution of knowledge items and their relations so as to assess the extent to which a new unit of technological knowledge feeds the generation of technological knowledge in other fields and the extent to which the generation of new technological knowledge in a field depends on the contributions of knowledge inputs from other fields.[4]

The generation of knowledge is enhanced by the selective recombination of ideas centred upon a set of core technologies with high levels of fungibility, and feeds the generation of further innovations by stimulating their knowledge compositeness. Gradually diminishing returns to recombination will limit the growth of new technological systems: excess variety matters. The introduction and dissemination of new ICTs in the last two decades of the twentieth century is characterized by this dynamic.

Detailed analysis of the characteristics of the knowledge base, drawing on patent statistics, enables us to identify the actual dynamics of recombinant knowledge by exploiting the distribution of patents across technological classes. We assume that the distribution of co-occurrences of technological classes across the patent portfolios of agents and countries can be considered a reliable indicator of the extent to which recombination is involved and has contributed to economic growth in each context.

The implementation of the indicators proxying the properties of the knowledge base is carried out by using patent statistics.[5] Note that, to introduce some rigidities into national technological portfolios and to compensate for the intrinsic volatility of patenting behaviour, each patent is assumed to be in force for five years. We calculated most of the relevant variables, like revealed technology advantage, technological variety and knowledge coherence, by relying on the technological classes assigned to each patent on the basis of the International Patent Classification (IPC).[6] Besides the *variety* and *coherence* indexes described in Section 5.2, we implemented also the following regressors:

1 First, the ICT knowledge stock is a proxy measure for the rate at which knowledge is produced within each country's ICT sector, traditionally used to measure the output from knowledge generating activities. It is computed for each country, at each year, by applying the permanent inventory method to patent applications. We calculate it as the cumulated stock of patent applications in the ICT field using a rate of obsolescence of 15 per cent per annum: $E_{s,i,t}=\dot{h}_{s,i,t}+(1-\delta)E_{s,i,t-1}$, where $\dot{h}_{s,i,t}$ is the flow of patent applications in sector s in country i, and δ is the rate of obsolescence.[7] This measure has some shortcomings, however, in that it is affected by cross-country size differences, which means we need an index able to discount for country size. To this end, it is useful to look at the ratio between ICT knowledge stock and total knowledge stock for each country at each year:

$$ICTK_{s,i,t} = \frac{E_{s,i,t}}{\sum_{s} E_{s,i,t}} \tag{6.2}$$

However, an index that is better suited to measuring the relative technological strengths (or weaknesses) of countries is represented by revealed technological advantage (RTA), developed by Soete (1987). This is defined as follows:

$$RTA_{s,i,t} = \left(\frac{E_{s,i,t}}{\sum_{s} E_{s,i,t}} \right) \Big/ \left(\frac{\sum_{i} E_{s,i,t}}{\sum_{s}\sum_{i} E_{s,i,t}} \right) \tag{6.3}$$

The RTA index varies around unity, such that values greater than 1 indicate that country i is relatively strong in technology s, compared to other countries and the same technological field, while values less than 1 indicate a relative weakness.[8]

2 As argued in Section 5.2, traditional measures of innovation built on a purely quantitative account of knowledge capital stock or qualitative indices based on patent citations and litigation do not capture the effects of variety, selective recombination and complementarity in the generation of technological knowledge. Thus, we use indices based on the co-occurrence of technological classes within patent applications. This means that the main focus of our analysis is on multi-technology patents, making it necessary to control for their time evolution by including the following variable in the regression:

$$MTP_{i,t} = \frac{\sum_{q}\sum_{i} E_{q,i,t}}{\sum_{i} E_{i,t}} \tag{6.4}$$

If q is the set of multi-technology patents, the index *MTP* in Equation (6.4) defines the share of these patents in the whole technological portfolio of each country in the ICT sector. It should be noted that the distribution of this variable is highly skewed to the right, as the knowledge stock in all the sampled countries is dominated by multi-technology patents from the beginning of the period of our analysis.

We are now able to qualify our working hypotheses by giving them an operational translation. In this chapter we hypothesize that the evolution of the knowledge base underlying ICTs is likely to trigger economic growth as long as it is articulated around a wide array of diverse but highly complementary technologies, while the concentration of emergent variety within well defined boundaries is likely to yield negative effects on technological opportunities and, hence, on economic growth.

More specifically we test the hypothesis that the amount of technological change introduced in an economic system, as measured by total factor productivity growth, will be larger:

a the larger the technological specialization of the knowledge activities within the system;

b the larger the coherence of the knowledge activities that take place within an economic system;

c the lower the related and unrelated variety of knowledge activities.

To test this hypothesis econometrically requires us to rewrite Equation (6.1) so as to model the MFP growth rate as a function of the knowledge base characteristics:

$$\ln\left(\frac{A_i(t)}{A_i(t-1)}\right) = a + b\ln A_{i,t-1} + c_1 \ln RTA_{i,t-1} + c_2 \ln R_{i,t-1} + c_3 \ln(TV_{i,t-1}) + \quad (6.5)$$
$$e\ln MTP_{i,t-1} + \sum_{i=1}^{15} \mu_i D_{i,t} + \sum_{t=1981}^{2003} \mu_t D_{i,t} + \varepsilon_{i,t}$$

The second part of Equation (6.5) comprises the control variables, where μ_i represents country-fixed effects, μ_t captures time-fixed effects and $MTP_{i,t-1}$ refers to the share of multi-technology patents in the ICT sector in each country. The first part of the equation represents the properties of the knowledge base, i.e. revealed technology advantage (*RTA*), knowledge coherence (*R*) and total variety index (*TV*). In order to appreciate the effects of related (RTV) and unrelated (UTV), we estimate Equation (6.5) alternating the three indexes for variety.

6.3 The data

In order to test the working hypothesis proposed in Section 5.3, we combine a dataset containing information on the economic variables with a dataset of patent applications. The former is used to calculate the MFP index described above. For this purpose we exploit the data on gross domestic product (GDP), labour income, employment and gross fixed capital formation from the OECD Stan database; information on total hours worked is taken from the Groningen Growth and Development Centre (www.ggdc.net).

Data on patent applications are drawn from the European Patent Office (EPO)[9] dataset (Espacenet). The identification of ICT-related patents is somewhat controversial, due to the criteria used to build the classifications. In particular, the use of the IPC has been criticized for its inherently function-oriented nature (Corrocher *et al.*, 2007). However, several empirical contributions use IPC to identify the borders of the ICT sector. We decided to merge the classification proposed by the OECD with those developed by the French *Observatoire des Sciences et des Techniques* (OST), in order to achieve a more inclusive representation. These classes are reported in Table 6.1.

The initial EPO dataset consisted of 115,771 patent applications, which we assigned to countries based on the first two digits of their priority number.[10] This allowed us to classify about 90 per cent of the dataset. The time coverage of the dataset was from 1978 to 2006: we focus on the period 1981–2003, and include only countries with observations for at least 22 years. The resulting sample includes 96,149 patent applications, distributed across 14 OECD countries.

Table 6.2 presents the dataset showing that the distribution of patent applications in the ICT field is rather skewed, with 42 per cent concentrated in the US. It should be noted that this is a considerable underestimation of the US weight; it would be reasonable to expect that US firms will tend to have more patents registered with the US Patent and Trademark Office (USPTO) than with the EPO. This also applies to Japanese patent applications, which in our case are 15 per cent of the observed total. In sum, 80 per cent of the patents in the telecommunication industry are concentrated in four countries, i.e. the US, Japan, Germany and France, with the UK ranked fifth with a share of about 7 per cent of total patent applications.

A look at the evolution of patenting in the ICT sector across countries confirms this preliminary evidence. Table 6.3 and Figure 6.1 report the breakdown of patent applications by country, cumulated over four years, to allow for the high degree of volatility of patent applications.

We can see that the gap between the US and the other countries analysed began to widen in the early 1990s (in Figure 6.1 US data are on the right y-axis). Japan's patent applications are initially below German and French applications: Japan overtakes France in 1994 and Germany in 2000. Note also that in the earliest years France is ranked higher than Germany and the UK, but was overtaken by Germany in 1995 and by the UK in 2000. We now turn to a detailed analysis of the dynamics of the properties of the ICT knowledge base in the sampled countries, in the context of the stylized facts on the evolution of the ICT sector.

Figure 6.2 depicts the aggregate dynamics of the core technological classes over time. In the first decade of our analysis there are two groups, based on frequency of technological classes. Most classes are cited in fewer than 100 patents in the period 1981–1986, and patent applications appear to be concentrated in a four classes, i.e. H03K (pulse technique), H04B (transmission), H04L (transmission of digital information) and H04Q (selecting). It is interesting that the first two classes, which are related to the communication aspect of ICTs, are the most frequent, while the last two, which are related more to the transmission of data in digital formats, although important are less developed.

From a dynamic viewpoint, the H04B class gained momentum in the early 1990s and continued sustained growth to 2003. The H04Q class followed

Table 6.1 IPC classes used to define the ICT sector

G08C	Transmission systems for measured values, control or similar signals
H01P	Waveguides; resonators, lines or other devices of the waveguide type
H01Q	Aerials
H03B	Generation of oscillations, directly or by frequency changing, by circuits employing active elements which operate in a non-switching manner; generation of noise by such circuits
H03C	Modulation
H03D	Demodulation or transference of modulation from one carrier to another
H03H	Impedance networks, e.g. resonant circuits; resonators
H03K	Pulse technique
H03L	Automatic control, starting, synchronization or stabilization of generators of electronic oscillations or pulses
H03M	Coding, decoding or code conversion, in general
H04B	Transmission
H04H	Broadcast communication
H04J	Multiplex communication
H04K	Secret communication; jamming of communication
H04L	Transmission of digital information, e.g. telegraphic communication
H04Q	Selecting

roughly the same path, although it remained at lower levels in absolute terms. The dynamics of H04L and H04J are also interesting. The former starts to increase at a fairly rapid rate after 1995, and from 1999 onwards is the class most frequently cited in patent applications. This is in line with anecdotal evidence that the convergence of computing and telecommunications technology became central in the 1990s, and 1995 corresponds roughly to the period of

Table 6.2 Cross-country distribution of patent applications

Country	*Freq.*	*%*	*Cum.*
US	41,963	43.64	43.64
JP	14,539	15.12	58.76
DE	10,867	11.3	70.06
FR	8,606	8.95	79.01
GB	7,420	7.72	86.73
SE	4,024	4.19	90.92
FI	3,806	3.96	94.88
NL	1,030	1.07	95.95
AU	974	1.01	96.96
IT	820	0.85	97.81
CH	667	0.69	98.5
CA	453	0.47	98.97
AT	339	0.35	99.32
NO	283	0.29	99.61
DK	266	0.28	99.89
BE	92	0.1	100
Total	**96,149**	**100**	

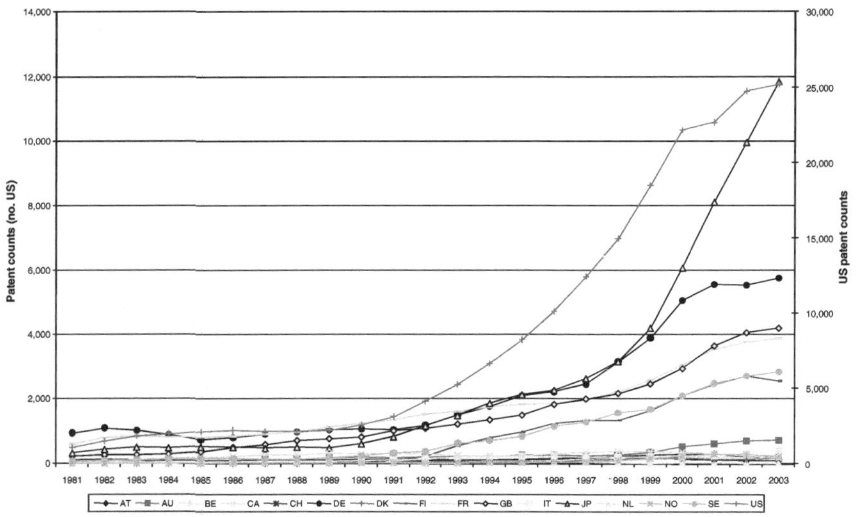

Figure 6.1 Patent applications in the ICT sector, four years' cumulative count.

Table 6.3 Country breakdown of patent applications in the ICT sector (four years cumulated), by year

	1981	*1982*	*1983*	*1984*	*1985*	*1986*	*1987*	*1988*	*1989*	*1990*	*1991*	*1992*	*1993*	*1994*	*1995*	*1996*	*1997*	*1998*	*1999*	*2000*	*2001*	*2002*	*2003*	*2004*	*2005*
AT	3	4	5	7	11	10	10	10	16	23	37	53	59	70	93	110	128	136	119	94	75	66	63	61	47
AU	22	27	30	35	40	51	63	79	92	105	102	99	115	112	133	137	122	134	181	265	314	366	365	323	269
CA	6	11	13	16	13	10	10	9	10	21	28	35	35	35	44	51	62	85	100	121	163	174	204	199	164
DE	530	625	590	517	437	484	564	625	677	684	672	749	949	1,138	1,352	1,471	1,650	2,115	2,644	3,420	3,785	3,736	3,826	3,476	2,839
DK	5	5	4	5	5	5	7	6	6	12	17	23	27	29	26	35	70	77	91	97	80	84	97	90	73
FR	386	521	576	544	540	545	578	629	750	832	909	1,031	1,063	1,164	1,226	1,243	1,338	1,542	1,809	2,131	2,458	2,626	2,713	2,563	2,071
GB	121	149	166	186	236	319	374	448	479	511	616	661	753	844	964	1,213	1,349	1,461	1,656	1,982	2,449	2,769	2,887	2,626	2,033
IT	33	39	34	27	22	31	39	46	48	61	75	97	135	141	149	152	142	145	151	153	177	236	295	306	263
JP	168	235	288	288	319	318	295	316	326	418	567	782	993	1,231	1,389	1,504	1,678	1,959	2,564	3,469	4,615	5,645	6,816	7,290	6,440
NL	35	55	79	95	119	140	170	185	203	211	210	195	162	165	171	207	235	240	240	235	215	185	144	92	61
NO	2	4	4	2	4	6	8	12	18	18	18	19	15	13	13	11	16	25	31	61	93	142	167	168	135
SE	43	52	82	95	107	118	97	98	119	163	219	243	376	420	496	713	836	1,050	1,161	1,399	1,635	1,708	1,750	1,407	908
US	606	843	1,093	1,225	1,302	1,382	1,332	1,361	1,440	1,732	2,110	2,762	3,426	4,249	5,230	6,420	7,995	9,725	12,043	14,437	14,615	15,782	16,138	13,997	12,579

Source: Elaborations on EPO data.

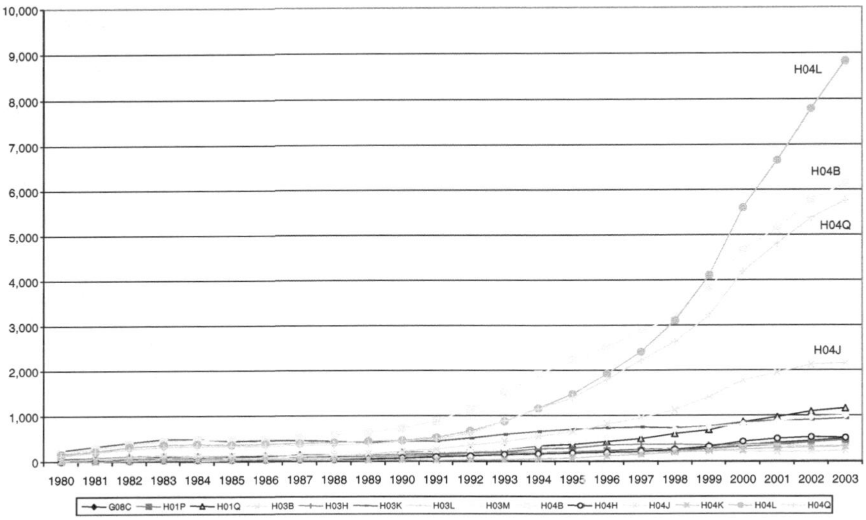

Figure 6.2 Dynamics of patent applications in the core ICT technological classes.

massive Internet diffusion and demonstration of its potential (van den Ende and Dolfsma, 2005; Fransman, 2007). The H04J class (multiplex communication) shows a marked increase in the late 1990s, corresponding with the surge in the technologies allowing for fast communication through the asynchronous transmission of digital signals on the existing infrastructures (such as ADSL).

6.4 Cross-country dynamics of ICT knowledge base: the empirical evidence

The evolution of ICTs and their diffusion within the economic system have had significant effects on economic performance, renewing productivity gaps between the US and other advanced countries. A large body of empirical literature documents this phenomenon, ascribing the success of the US economy up to the second half of the 1990s to the ability to trigger demand for ICTs, and the simultaneous rise of the services sector (Jorgenson, 2001).

The continuing US leadership in the ICT sector suggests the existence of a path of continuing exploitation of the technological opportunities uncovered by research in the field. This is the case at least until the early 1990s. The change in technological focus from the component to the content/application layer coincides with a marked discontinuity in technological competences. The parallel developments of the other advanced countries suggests that those with a relatively strong commitment to research in the ICT sectors have been able, through imitation, to follow the US along this technological path. At the same time, countries with a weaker research focus have experienced somewhat less favourable dynamics.

It is important, therefore, to explore the evolution of the relative intensity of research in the sampled countries. Table 6.4 reports the dynamics of RTA, calculated according to Equation (6.2). The results of our calculations show that our sample of OECD countries falls roughly into three groups, according to the actual levels of RTA and its dynamics:

1 First, there is a large number of older competitors or the incumbents (including the US, the UK, France, Germany and Australia), which are characterized by relatively high levels of RTA already in the 1980s. Most are characterized by increasing RTA in the 1980s followed by a decrease in the 1990s. The US is an exception in that its RTA in ICT increases continuously during the 1980s and the 1990s, and at an even rate.
2 Second, there is the group of late-leading countries, which includes a few north European countries, mainly Finland, Norway and Sweden. These countries are characterized by low levels of RTA in the 1980s (especially in Finland and Norway) and a steep increase in RTA in the 1990s, allowing them to overcome the group of incumbents.
3 Third, there is the group of laggards, such as Canada, Japan, Italy, etc. These countries exhibit quite low levels of RTA, and it is difficult to identify any pattern of evolution. For example, the RTA index is continuously increasing in the case of Japan, while it is stable for Canada and constantly decreasing for Italy.

This grouping has some interesting implications in terms of variety indexes. Table 6.5 reports the breakdown by country of the evolution of general (or total) variety. It is evident that the incumbent countries (the first group) are characterized by the highest levels of the variety index. The dynamics are generally quite stable over time, with the exception of Australia, whose variety index rapidly increased in the 1980s, reaching the same levels as the other countries in the group. Out of the late-leaders, the technological variety index for Sweden increases smoothly during the 1980s, remaining stable in the 1990s at levels very similar to the incumbent countries. The dynamics for Finland and Norway are characterized by a marked increase in the 1980s, and a table pattern along the 1990s at levels lower than for Sweden. Finally, the index of variety for the group of lagging countries shows no clear-cut pattern. Japan's is similar to the incumbent countries, while Austria and Canada are characterized by low levels in the 1980s which increase rapidly in the 1990s.

The general variety index can be decomposed into related (Table 6.6) and unrelated (Table 6.7) variety, both tables showing that the incumbent group of countries is characterized by high levels of related variety, mostly stable over time, with unrelated variety generally at lower levels across the time span. Late-leading countries generally have high and increasing levels of related variety (though generally below the values for incumbents), and especially in the 1990s, and comparatively low levels of unrelated variety, although in the case of Norway and Finland in the 1980s, unrelated variety has a higher weight than

Table 6.4 Country breakdown of revealed technology advantage in the ICT sector

	AT	*AU*	*BE*	*CA*	*DE*	*DK*	*FR*	*GB*	*IT*	*JP*	*NL*	*NO*	*SE*	*US*
1981	0.139	1.529	0.122	0.284	1.071	0.381	1.582	0.714	0.494	0.687	0.539	0.190	0.801	1.089
1982	0.158	1.391	0.135	0.351	0.967	0.276	1.684	0.678	0.481	0.688	0.682	0.611	0.803	1.138
1983	0.150	1.236	0.257	0.353	0.859	0.274	1.671	0.678	0.368	0.695	0.790	0.485	1.108	1.216
1984	0.154	1.381	0.301	0.314	0.805	0.339	1.715	0.741	0.292	0.623	0.906	0.336	1.163	1.292
1985	0.209	1.580	0.339	0.275	0.764	0.324	1.724	0.880	0.269	0.638	1.126	0.500	1.137	1.298
1986	0.175	1.708	0.297	0.239	0.758	0.253	1.753	1.054	0.323	0.583	1.213	0.840	1.161	1.321
1987	0.161	1.888	0.445	0.226	0.780	0.369	1.808	1.138	0.330	0.532	1.403	0.801	1.228	1.303
1988	0.172	2.350	0.498	0.258	0.781	0.286	1.858	1.277	0.320	0.505	1.390	0.823	1.329	1.304
1989	0.384	3.101	0.391	0.263	0.771	0.272	1.975	1.356	0.308	0.457	1.509	1.081	1.553	1.277
1990	0.389	3.210	0.347	0.411	0.730	0.408	1.967	1.427	0.375	0.466	1.501	1.236	1.840	1.313
1991	0.469	2.726	0.283	0.468	0.683	0.591	1.878	1.569	0.374	0.508	1.476	1.009	2.068	1.325
1992	0.543	2.560	0.234	0.484	0.649	0.543	1.796	1.492	0.377	0.561	1.205	0.882	1.871	1.367
1993	0.558	2.675	0.181	0.368	0.655	0.533	1.628	1.412	0.408	0.555	0.975	0.811	2.496	1.344
1994	0.539	2.255	0.337	0.371	0.616	0.524	1.516	1.365	0.374	0.588	0.869	0.634	2.264	1.395
1995	0.650	2.151	0.260	0.372	0.600	0.419	1.401	1.404	0.346	0.588	0.818	0.479	2.228	1.438
1996	0.700	1.965	0.215	0.350	0.543	0.399	1.287	1.451	0.303	0.563	0.744	0.388	2.380	1.520
1997	0.649	1.743	0.152	0.296	0.533	0.688	1.182	1.362	0.269	0.563	0.639	0.454	2.399	1.562
1998	0.583	1.540	0.120	0.320	0.553	0.593	1.150	1.266	0.232	0.591	0.564	0.496	2.324	1.587
1999	0.476	1.548	0.083	0.282	0.547	0.521	1.084	1.228	0.207	0.635	0.477	0.399	2.054	1.624
2000	0.364	1.602	0.077	0.268	0.544	0.437	1.026	1.233	0.175	0.698	0.376	0.619	2.195	1.617
2001	0.307	1.528	0.060	0.285	0.525	0.402	1.034	1.337	0.170	0.797	0.297	0.858	2.375	1.502
2002	0.267	1.463	0.047	0.284	0.492	0.381	1.011	1.314	0.186	0.857	0.242	1.090	2.370	1.516
2003	0.239	1.420	0.038	0.313	0.494	0.406	0.972	1.293	0.195	0.933	0.200	1.133	2.238	1.508

Source: Elaborations on EPO data.

Table 6.5 Country breakdown of variety (information entropy) in the ICT sector

	AT	*AU*	*BE*	*CA*	*DE*	*DK*	*FR*	*GB*	*IT*	*JP*	*NL*	*NO*	*SE*	*US*
1981	1.585	4.265	0.000	1.000	7.025	0.918	6.887	6.520	3.785	6.677	3.546		4.999	7.333
1982	2.000	4.472	0.000	0.918	7.237	0.918	7.077	6.529	4.415	6.976	4.537	1.500	5.238	7.486
1983	2.000	4.963	1.500	0.722	7.325	0.918	7.171	6.702	4.415	7.084	5.044	1.500	5.319	7.541
1984	2.000	5.127	2.522	0.811	7.212	2.750	7.250	6.551	4.252	6.900	5.502	1.500	4.793	7.591
1985	2.585	5.330	2.752	0.811	7.238	2.750	7.290	6.257	4.022	6.905	5.846	1.918	4.681	7.603
1986	2.322	5.250	2.689	0.000	7.242	2.585	7.412	6.377	4.133	6.770	6.009	2.752	4.742	7.700
1987	2.322	5.446	3.071	0.000	7.130	2.750	7.462	6.494	3.759	6.760	6.092	2.000	5.122	7.697
1988	1.922	5.503	3.201	1.585	7.175	2.750	7.597	6.464	4.101	6.884	6.035	2.000	5.209	7.742
1989	3.476	6.246	3.190	2.250	7.096	0.000	7.558	6.508	4.324	6.876	6.179	2.522	5.519	7.657
1990	3.372	6.290	3.182	3.093	7.258	2.948	7.681	6.498	4.751	6.998	6.265	3.932	5.688	7.701
1991	3.877	6.342	2.664	3.484	7.278	3.922	7.637	6.617	5.202	7.029	6.334	3.807	5.763	7.667
1992	4.564	6.599	2.252	3.546	7.393	4.005	7.576	6.777	5.511	7.151	6.377	3.875	5.766	7.786
1993	4.750	6.778	1.500	3.427	7.368	4.670	7.470	6.864	5.639	7.139	6.313	4.022	6.052	7.829
1994	4.887	6.355	4.004	3.793	7.608	4.960	7.429	6.989	5.589	7.196	6.170	3.932	5.978	7.890
1995	5.099	6.496	3.924	3.446	7.647	4.626	7.189	7.069	5.505	7.089	5.949	2.250	6.137	7.963
1996	5.294	6.459	4.180	3.805	7.578	4.317	7.153	7.024	5.342	7.139	5.666	2.250	6.228	7.998
1997	5.124	6.436	4.180	3.792	7.605	5.226	7.190	6.855	5.086	7.524	5.698	2.896	6.200	7.927
1998	5.266	6.282	4.378	3.954	7.507	5.217	7.112	6.717	4.788	7.604	5.955	2.583	6.044	7.777
1999	5.214	6.553	2.918	3.638	7.381	5.282	7.018	6.741	5.052	7.454	5.793	2.422	6.247	7.658
2000	5.441	6.684	3.250	4.063	7.272	5.348	7.057	6.670	4.918	7.509	5.828	3.274	6.222	7.571
2001	5.138	6.843	2.722	3.867	7.364	5.362	6.997	6.609	4.972	7.419	5.714	3.515	6.253	7.571
2002	5.083	6.852	2.722	4.196	7.383	5.288	6.954	6.642	5.357	7.375	5.311	3.572	6.435	7.522
2003	4.536	7.019	1.585	4.859	7.581	5.589	6.945	6.607	5.575	7.443	4.881	4.218	6.264	7.461

Source: Elaborations on EPO data.

Table 6.6 Country breakdown of related variety (within-group information entropy) in the ICT sector

	AT	*AU*	*BE*	*CA*	*DE*	*DK*	*FR*	*GB*	*IT*	*JP*	*NL*	*NO*	*SE*	*US*
1981	0.000	2.852	0.000	0.000	5.488	0.000	4.967	4.158	2.619	5.003	1.968		3.049	5.367
1982	0.000	2.986	0.000	0.000	5.481	0.000	5.161	4.261	3.306	5.278	3.080	0.500	3.417	5.578
1983	0.500	3.557	0.689	0.000	5.510	0.000	5.277	4.595	3.279	5.222	3.628	0.500	3.574	5.657
1984	0.500	3.700	1.641	0.000	5.447	0.844	5.319	4.562	3.076	5.113	4.181	0.500	3.326	5.740
1985	1.126	3.795	1.833	0.000	5.258	0.844	5.273	4.503	3.051	5.123	4.554	0.333	3.365	5.816
1986	1.351	3.443	1.853	0.000	5.105	0.667	5.578	4.457	3.150	4.971	4.574	0.167	3.448	5.829
1987	1.351	3.659	2.349	0.000	5.101	0.500	5.526	4.618	2.668	4.928	4.703	0.000	3.529	5.863
1988	0.951	3.713	2.747	1.585	5.141	0.500	5.647	4.503	2.233	5.245	4.733	0.000	3.319	5.860
1989	2.723	4.131	2.623	1.439	5.181	0.000	5.527	4.652	2.352	5.215	4.625	0.000	3.643	5.877
1990	2.981	4.266	2.093	2.230	5.325	2.948	5.681	4.725	2.901	5.370	4.858	1.851	3.858	5.992
1991	3.354	4.354	1.386	2.550	5.335	2.516	5.575	5.016	3.394	5.339	4.964	2.807	4.084	5.976
1992	3.393	4.465	0.459	2.813	5.516	2.653	5.607	5.131	3.718	5.645	4.958	2.886	4.298	6.085
1993	3.737	4.532	0.000	2.643	5.399	2.809	5.545	5.228	4.156	5.689	4.818	3.029	4.231	6.195
1994	3.646	4.229	1.630	2.641	5.463	3.269	5.625	5.266	4.278	5.804	4.811	2.934	4.113	6.195
1995	3.902	4.246	1.809	2.431	5.495	2.826	5.553	5.381	3.890	5.674	4.376	1.250	4.130	6.309
1996	3.601	4.113	1.710	2.109	5.491	2.448	5.650	5.315	3.734	5.780	4.149	1.296	4.316	6.286
1997	3.568	4.186	1.710	1.885	5.357	3.403	5.713	5.247	3.468	5.432	4.283	1.956	4.338	6.245
1998	3.643	4.167	1.750	1.987	5.404	3.337	5.653	5.224	3.163	5.572	4.445	1.664	4.524	6.143
1999	3.700	4.667	0.801	1.933	5.386	3.340	5.618	5.225	3.601	5.485	4.338	1.541	4.647	6.070
2000	3.449	4.609	0.892	2.204	5.375	3.525	5.557	4.974	3.879	5.628	4.209	2.021	4.895	5.941
2001	3.413	4.751	1.151	2.847	5.391	3.647	5.403	4.912	3.839	5.560	3.872	2.302	4.884	5.917
2002	3.125	4.764	1.151	3.236	5.405	3.605	5.294	4.906	3.991	5.616	3.220	2.406	4.944	5.856
2003	2.428	4.840	0.667	3.657	5.460	4.307	5.254	4.848	3.994	5.648	2.719	2.833	4.743	5.788

Source: Elaborations on EPO data.

Table 6.7 Country breakdown of unrelated variety (between-group information entropy) in the ICT sector

	AT	*AU*	*BE*	*CA*	*DE*	*DK*	*FR*	*GB*	*IT*	*JP*	*NL*	*NO*	*SE*	*US*
1981	1.585	1.412	0.000	1.000	1.537	0.918	1.920	2.362	1.166	1.674	1.578		1.950	1.966
1982	2.000	1.487	0.000	0.918	1.755	0.918	1.916	2.268	1.109	1.698	1.457	1.000	1.821	1.908
1983	1.500	1.406	0.811	0.722	1.815	0.918	1.893	2.107	1.136	1.862	1.417	1.000	1.745	1.884
1984	1.500	1.427	0.881	0.811	1.765	1.906	1.931	1.989	1.175	1.786	1.321	1.000	1.467	1.851
1985	1.459	1.535	0.918	0.811	1.980	1.906	2.018	1.754	0.971	1.782	1.292	1.585	1.316	1.787
1986	0.971	1.807	0.837	0.000	2.136	1.918	1.834	1.920	0.983	1.799	1.435	2.585	1.294	1.871
1987	0.971	1.786	0.722	0.000	2.029	2.250	1.936	1.876	1.091	1.832	1.389	2.000	1.593	1.834
1988	0.971	1.790	0.454	0.000	2.033	2.250	1.950	1.961	1.867	1.639	1.302	2.000	1.891	1.883
1989	0.753	2.115	0.567	0.811	1.915	0.000	2.031	1.856	1.972	1.662	1.554	2.522	1.875	1.781
1990	0.391	2.024	1.089	0.863	1.933	0.000	2.001	1.773	1.849	1.628	1.407	2.082	1.829	1.709
1991	0.523	1.987	1.278	0.934	1.943	1.406	2.061	1.601	1.808	1.690	1.370	1.000	1.679	1.691
1992	1.170	2.134	1.792	0.734	1.876	1.352	1.968	1.646	1.794	1.505	1.419	0.989	1.469	1.700
1993	1.014	2.246	1.500	0.784	1.969	1.861	1.925	1.636	1.483	1.450	1.495	0.993	1.822	1.633
1994	1.241	2.125	2.374	1.152	2.146	1.691	1.804	1.722	1.311	1.392	1.359	0.998	1.865	1.695
1995	1.197	2.251	2.115	1.015	2.151	1.800	1.637	1.689	1.615	1.415	1.572	1.000	2.006	1.653
1996	1.693	2.346	2.470	1.695	2.087	1.870	1.503	1.709	1.608	1.359	1.517	0.954	1.911	1.712
1997	1.556	2.250	2.470	1.907	2.247	1.824	1.477	1.607	1.618	2.092	1.415	0.940	1.862	1.682
1998	1.624	2.115	2.628	1.967	2.103	1.879	1.459	1.493	1.625	2.032	1.509	0.918	1.520	1.634
1999	1.514	1.886	2.117	1.705	1.994	1.942	1.399	1.516	1.451	1.969	1.455	0.881	1.599	1.588
2000	1.992	2.075	2.358	1.859	1.897	1.823	1.500	1.695	1.039	1.881	1.618	1.253	1.327	1.630
2001	1.725	2.092	1.571	1.020	1.973	1.715	1.594	1.697	1.133	1.860	1.842	1.213	1.369	1.653
2002	1.957	2.087	1.571	0.960	1.978	1.683	1.660	1.736	1.366	1.759	2.090	1.166	1.490	1.666
2003	2.107	2.179	0.918	1.201	2.120	1.283	1.692	1.759	1.582	1.795	2.162	1.385	1.522	1.673

Source: Elaborations on EPO data.

related variety. In the laggard group, the dynamics for Japan are similar to those of the incumbents, while for Canada, Austria and Italy, unrelated variety has a comparatively higher weight in the 1980s, and lower weight in the 1990s.

The evidence on RTA and variety is reflected in the dynamics of knowledge coherence (Table 6.8). US values for knowledge coherence are positive in the first half of the 1980s when research in the ICT sector was focused on the component level, and was exploiting the technological potentials established in the 1960s and 1970s. The emergence of the technical conditions leading to Internet diffusion, and the related shift in technological efforts towards the development of content applications, introduced a discontinuity that is reflected in the falling coherence index along the 1990s.

Within the group of incumbents, France shows increasing coherence along the 1980s with a positive index in 1984, which then dropped to below zero in the 1990s. The values for Germany during the 1980s fluctuate around zero, being negative until 1983 and then positive up to 1992, and negative for the remainder of the 1990s. The countries in the other two groups are also characterized by dramatic falls in knowledge coherence during the observed period. The evidence for Canada is noteworthy in that in the early 1980s the index is quite high, but decreases over time and in 2003 is lower than any other sampled country.

6.5 Econometric results

In order to assess the effects of the properties of the knowledge base on MFP, we carried out panel data fixed-effects estimations of Equation (6.5). The results are reported in Table 6.9 and Table 6.10. The estimations differ in that in the former we proxied the relative weight of ICTs in each country by the ratio between ICT knowledge stock and total knowledge stock, following Equation (6.1). In the latter (Table 6.10) we use the RTA index, which gives us information on the relative technological specialization of each country in the ICT sector.

Table 6.9 column (1) reports the estimation by considering total variety. The coefficient of the share of knowledge stock produced in the ICT sectors has a positive and significant sign. As expected, productivity growth is likely to grow as the share of ICT-related knowledge increases. The coefficient of knowledge coherence is also positive and significant. Again, consistent with our working hypotheses, the clustering of knowledge generating activities around a distinctive core of technologies is likely to enhance the innovation process and trigger productivity growth. The higher the degree of internal coherence of the knowledge base, the better the economic performance.

The negative and significant sign for variety is also in line with our theoretical framework and does not contradict existing firm- and regional-level evidence (Nesta, 2008; Quatraro, 2010). Our results contrast with the findings of recent empirical studies on the effects of technological diversity on firms' innovative performance, which show positive and significant coefficients (Nesta and Saviotti, 2005; Leten *et al.*, 2007; Garcia-Vega, 2006; D'Este, 2005). However, we

Table 6.8 Country breakdown of knowledge coherence in the ICT sector

	AT	*AU*	*BE*	*CA*	*DE*	*DK*	*FR*	*GB*	*IT*	*JP*	*NL*	*NO*	*SE*	*US*
1981	–0.438	–1.822	–1.578	5.447	–0.125	–0.909	–0.135	–0.218	–1.726	0.502	–1.468		–0.860	0.367
1982	–2.219	–1.910	–3.201	3.881	–0.185	–0.686	–0.158	–0.363	–0.931	0.252	–0.846	–2.315	–1.470	0.361
1983	–1.828	–1.192	–6.432	1.743	–0.153	–1.993	–0.064	–0.435	–1.576	0.573	–0.940	11.333	–1.555	0.261
1984	–2.643	–1.690	–4.307	2.239	0.174	–1.481	0.133	–0.305	–1.141	0.457	–0.868	–0.242	–1.097	0.412
1985	–2.086	–1.195	–3.983	0.605	0.103	0.128	0.342	–0.040	–1.495	0.447	–0.598	5.888	–1.094	0.377
1986	–2.501	–0.930	0.208	0.811	0.067	0.427	0.443	0.007	–0.834	0.116	–0.462	6.956	–1.407	0.351
1987	–2.049	–0.664	–1.403	0.074	0.010	–1.800	0.239	–0.012	–1.108	0.170	–0.408	–0.760	–0.784	0.198
1988	–3.005	–1.145	–0.863	0.809	–0.071	–2.350	0.255	–0.074	–1.358	0.318	–0.519	0.949	–1.247	0.502
1989	–1.953	–0.787	0.204	–0.079	0.114	–4.207	0.453	–0.123	–1.202	0.075	–0.176	2.149	–0.967	0.297
1990	–1.011	–1.147	–0.540	–2.412	0.092	–0.031	0.305	–0.259	–1.900	–0.156	–0.246	0.188	–1.252	0.075
1991	–0.818	–1.675	–1.514	–1.254	0.134	–0.661	0.227	–0.407	–1.753	–0.300	–0.491	–0.028	–1.241	–0.025
1992	–1.011	–1.860	–1.828	–3.364	0.064	–1.372	–0.036	–0.456	–1.309	–0.223	–0.430	–0.922	–1.683	–0.125
1993	–2.568	–1.671	–3.072	–2.469	–0.079	–2.071	–0.169	–0.714	–1.704	–0.369	–0.987	–1.693	–1.903	–0.410
1994	–2.909	–1.924	–3.557	–2.479	–0.414	–3.733	–0.613	–1.031	–1.904	–0.569	–1.634	–1.566	–2.304	–0.605
1995	–2.224	–1.870	–2.885	–2.955	–0.455	–3.601	–0.819	–1.297	–2.395	–0.597	–1.893	–1.882	–2.331	–0.784
1996	–2.907	–1.446	–2.900	–4.499	–0.677	–2.231	–1.073	–1.727	–2.567	–0.738	–2.500	–2.020	–2.560	–0.879
1997	–3.611	–2.706	–3.280	–4.375	–0.964	–2.968	–1.303	–1.959	–2.453	–0.827	–2.686	–3.175	–2.706	–1.060
1998	–3.980	–2.301	–3.370	–6.027	–1.425	–1.167	–1.547	–2.464	–2.564	–0.979	–2.932	–6.298	–2.889	–1.375
1999	–4.770	–3.041	–3.984	–6.630	–1.713	–3.602	–1.882	–2.750	–3.060	–1.241	–3.463	–6.720	–3.066	–1.646
2000	–4.990	–3.402	–4.473	–6.819	–1.952	–3.784	–2.292	–3.007	–3.268	–1.650	–3.540	–6.540	–3.564	–1.907
2001	–5.063	–3.656	–4.107	–7.978	–2.212	–4.313	–2.681	–3.482	–3.875	–1.944	–3.968	–7.987	–3.288	–2.096
2002	–4.527	–3.518	–4.426	–7.789	–2.329	–5.351	–2.887	–3.238	–4.750	–1.987	–4.240	–6.963	–3.379	–2.172
2003	–5.764	–3.563	–2.292	–8.622	–2.602	–5.629	–3.050	–3.494	–5.209	–2.163	–4.121	–5.712	–3.556	–2.287

Source: Elaborations on EPO data.

cannot compare the findings from these studies with the present analysis for a number of reasons. First, most of these studies focus on the effects of technological diversification on innovation performance, using patent numbers as a dependent variable. It would be expected that an increase in patents will be accompanied by an increase in technological diversity (and vice versa). However, this does not necessarily apply to productivity, which measures the extent to which profitable innovations have been successfully adopted by economic agents. Moreover, technological diversity is proxied either by the inverse Herfindahl index or by a measure of technological scope, which is different from measuring technological variety based on information entropy. We should also add that all the studies referred to above consider the occurrence of a single technological class and not combinations of technological classes, whereas our study investigates the effects on productivity growth of technological variety captured by the overlapping of technological classes as measured by the co-occurrence of technological classes within the same patent. The use of multidimensional information entropy allows us to quantify the extent to which growth in technological activity is characterized by an increase in the observed combinations of technological classes (Saviotti, 1988).

Our results confirm that search processes directed towards new technological fields, leading to previously untried knowledge recombination, characterize the changes in the technological environment. During the early phases of this process, information entropy is likely to increase. Once the technological system is established, the technological environment becomes relatively stable. Establishment of the technological system is characterized by the likely introduction of incremental innovations within well defined technological boundaries.

During the mature stage of the technology lifecycle, innovation activities are likely to be directed towards the search for new applications of the knowledge base, featuring the particular technological system. These applications may well be outside the original technological boundaries, but may still be profitable, as in the case of the application of ICTs to the manufacture of medical devices, which is the same as our measure of unrelated variety. However, the increase in unrelated variety leads to an increased probability of less fertile combinations being explored. For this reason, at the aggregate level we would expect unrelated variety to have a negative effect on productivity growth. The opposite argument holds in the case of related variety, which is likely to characterize the establishment of the technological system and the phase of exploitation of its technological opportunities.

At a general level it is difficult, therefore, to predict the sign of the economic effects of technological variety, as they are largely influenced by the relative stage of development of the technological system under scrutiny, and by the associated dominance of related and unrelated variety. Diminishing returns to variety are likely to emerge in the mature stage when technological activities are featured by random screening across brand new combinations. As a consequence, when unrelated (related) variety shapes the evolution of technological variety, this latter is likely to have a negative (positive) effect on economic performance (Krafft *et al.*, 2009).

Table 6.9 Econometric estimation of Equation (6.5)

	(1)	*(2)*	*(3)*	*(4)*
Constant	0.0221***	0.0221***	0.0206***	0.0238***
	(0.00507)	(0.00495)	(0.00503)	(0.00506)
lagA	0.00768	0.00650	0.00837	0.00713
	(0.00615)	(0.00613)	(0.00620)	(0.00612)
Coherence	0.00365**	0.00352**	0.00322*	0.00394**
	(0.00191)	(0.00178)	(0.00191)	(0.00189)
Technological variety	–0.00254*			
	(0.00153)			
Unrelated technological variety		–0.00229***		–0.00287***
		(0.000853)		(0.000931)
Related technological variety			–0.000473	–0.00255
			(0.00155)	(0.00167)
Share of multi-tech patents	0.000360	0.000624	7.14e–06	0.00109
	(0.00109)	(0.00108)	(0.00108)	(0.00112)
ICTK	0.00180*	0.000524	0.00133	0.00146
	(0.00118)	(0.00113)	(0.00130)	(0.00128)
Time dummies	Yes	Yes	Yes	Yes
Country dummies	Yes	Yes	Yes	Yes
Observations	315	315	315	315
Number of countries	14	14	14	14
R-squared	0.371	0.381	0.365	0.386

Notes
Dependent variable: dlogA/dt.
Standard errors in parentheses. *** $p < 0.01$, ** $p < 0.05$, * $p < 0.1$.

We need to understand which of these two factors is likely to drive total variety. In columns (2) and (3) of Table 6.9 the index is articulated respectively as unrelated and related variety. Nevertheless, the results seem consistent with our argument of diminishing returns to recombination. The econometric findings show that the effect of related variety on productivity growth is not statistically significant, while the coefficient of unrelated variety is negative and significant. This means that the observed negative effect of technological variety is driven by its "unrelated" component. This result is consistent with the evidence on knowledge coherence, which again has a positive and significant coefficient. The increase in knowledge coherence is likely to be associated with increasing productivity growth rates. When knowledge coherence increases, then unrelated variety will fall or related variety will increase, or both. Our results shows that the patterns of productivity growth are characterized by a decrease in unrelated variety and non-significant changes in related variety.

Table 6.10 presents the results for the estimations including the *RTA* instead of *ICTK*. The coefficient of the RTA is positive and significant (Column (1)). This amounts to saying that the degree of relative technological specialization of countries in ICT has a positive effect on productivity growth. Productivity gaps, therefore, may be ascribed in part to the different technological focus of countries. Knowledge coherence has a positive and significant sign, in line with the previous estimation and the general theoretical framework underpinning the analysis. Total variety index, again, is negatively related to MFP growth and in this case calls for a deeper understanding of the relative impact of related and unrelated variety.

Columns (2) and (3) respectively present the effects of unrelated and related variety. Overall, the results are very similar to the previous estimations. The positive and significant sign of knowledge coherence is persistent across models and estimations, confirming the robustness of this result, and the coefficients of related and unrelated variety are in line with the previous estimation. The negative effects of technological variety seem to be driven by unrelated variety: the coefficient is negative and significant. Related variety does not seem to have an appreciable effect on cross-country differential growth rates.

The results of our estimations provide support for the hypothesis that the generation of knowledge in the ICT sector is likely to trigger productivity growth due to the inherent general purpose character of the technology. ICTs emerged from the recombination of a number of distinct bits of knowledge, from different technological fields, but with high degrees of complementarity. Failure to bring together complementary knowledge is likely to result in reduced knowledge coherence and an increase in unrelated variety, both of which are detrimental to productivity growth.

Table 6.10 Econometric estimation of Equation (6.5)

	(1)	*(2)*	*(3)*	*(4)*
Constant	0.0240***	0.0226***	0.0218***	0.0251***
	(0.00510)	(0.00486)	(0.00508)	(0.00512)
lagA	0.00896	0.00706	0.00916	0.00806
	(0.00624)	(0.00623)	(0.00634)	(0.00625)
Coherence	0.00361**	0.00350**	0.00316*	0.00387**
	(0.00191)	(0.00176)	(0.00190)	(0.00189)
Tech variety	–0.00265*			
	(0.00155)			
Unrelated tech variety		–0.00229***		–0.00289***
		(0.000845)		(0.000931)
Related tech variety			–0.000316	–0.00242
			(0.00147)	(0.00160)
RTA	0.00159*	0.000597	0.00107	0.00122
	(0.000991)	(0.000926)	(0.00102)	(0.00101)
Share of multi-tech patents	0.000342	0.000584	–2.81e–05	0.00105
	(0.00109)	(0.00108)	(0.00108)	(0.00112)
Time dummies	Yes	Yes	Yes	Yes
Country dummies	Yes	Yes	Yes	Yes
Observations	315	315	315	315
Number of countries	14	14	14	14
R-squared	0.371	0.381	0.365	0.386

Notes
Dependent variable: dlogA/dt.
Standard errors in parentheses. *** $p < 0.01$, ** $p < 0.05$, * $p < 0.1$.

6.6 Conclusions

The dynamics of knowledge generation is a challenging area of investigation. According to a growing literature on the system dynamics of technological change, new knowledge emerges from the recombination of existing knowledge. The characteristics of the map into which the recombination process takes place are most important. Knowledge recombination is more effective and fertile when and where the different knowledge items available are characterized by lower levels of variety and higher levels of specialization and coherence. In these circumstances recombination takes place more effectively and it can lead to the introduction of a new technological system. Knowledge recombination in this case is a process whose onset is characterized by the convergence of a core of complementary technologies. The steps that follow are fuelled by the gradual spread of the core to a growing number of other knowledge fields. Eventually, diminishing returns to knowledge recombination emerge.

Analysis of the co-occurrences of technologies within patent stocks allows us to study empirically the dynamics of knowledge recombination. Co-occurrences can be considered a reliable indicator of the overlapping of a new knowledge across existing technological classes. Frequency is relevant: only a few patents fall within just one technological class. The distribution of these co-occurrences and their dynamics can reveal key information about the emergence of new core technologies and their eventual growth into technological systems. Representing the knowledge base as a network, with an emphasis on its dynamic aspects, enables the identification of the changing structure of technological knowledge.

In this chapter we applied this theoretical framework and related empirical methodology, to the ICT sector, for the period 1983–2003. ICTs have been a major source of new technological knowledge and technological innovations, and became the engine of economic growth in the advanced countries in the last two decades of the twentieth century and the first years of the twenty-first century.

The rich empirical evidence on the dynamics of technological knowledge derived from analysis of the co-occurrence of technological classes within patents issued by the EPO in the period 1981–2003, across the different classes, has enabled the identification of a clear sequence in the development of technological knowledge. Following a period of concentrated technological advance in a few patent classes, we identified a phase of sustained recombinant growth.

Systematic exploration of the knowledge base using measures such as related and unrelated variety, coherence and cognitive distance, confirm that the grafting of recombinant ICT knowledge onto an increasing array of other patent classes has characterized the growth of technological knowledge since the 1980s. The structure of the knowledge base varies across countries and over time. Based on our evidence, countries can be categorized in three groups. The first consists of the older incumbents and includes the US, the UK, France, Germany and Australia, which, already in the 1980s, were characterized by relatively high levels of knowledge stock. The second is a group of fast-leading countries including

Finland, Norway and Sweden, which are characterized by a low-level knowledge base in the 1980s but show a steep increase in the 1990s. The third group gathers together laggards such as Canada, Italy, etc.

Our dynamic network analysis of the evolution of knowledge co-occurrence in two or more patenting classes has identified a clear pattern of evolution of the knowledge base. The incumbent group was the first to undergo a process of branching out of ICT knowledge, and a sustained phase of recombinant growth of the knowledge base. Digital knowledge promoted the emergence of new technological knowledge in a wide variety of other technological fields. Other fast-moving countries have proved able to catch up to an extent, but the laggards have been excluded from the benefits of recombinant growth.

Our empirical results support the basic hypothesis that the evolution of the knowledge base underlying ICTs in the form of recombinant knowledge has favoured economic growth through the application of a new core of highly complementary technologies. Attempts to extend knowledge recombination efforts beyond well defined boundaries of strong complementarity show a decline in technological opportunities with negative effects on the rates of increase of MFP and, hence, economic growth. Countries best able to master recombinant dynamics have proven able to achieve more rapid increase of their MFP growth.

Appendix 6.A: multifactor productivity calculations

In order to investigate the effects of the characteristics of ICT knowledge base on productivity growth, we first calculate an index of multifactor productivity (MFP) following the standard growth accounting approach (Solow, 1957; Jorgenson, 1995; OECD, 2001). We start by assuming that the national economy can be represented by a general Cobb–Douglas production function with constant returns to scale:

$$Y_{it} = A_{it} C_{it}^{\alpha_{it}} L_{it}^{\beta_{it}} \tag{A.1}$$

where L_{it} is the total hours worked in country i at time t, C_{it} is the level of the capital stock in country i at time t, and A_{it} is the level of MFP in country i at time t.

Following Euler's theorem, output elasticities are calculated (not estimated) using accounting data, assuming constant returns to scale and perfect competition in both product and factor markets.[11] The output elasticity of labour therefore is computed as the factor share in total income:

$$\beta_{i,t} = (w_{i,t} L_{i,t}) / Y_{i,t} \tag{A.2}$$

$$\alpha_{i,t} = 1 - \beta_{i,t} \tag{A.3}$$

where w is the average wage rate in country i at time t. Thus, we obtain elasticities that vary both over time and across countries.

The discrete approximation of the annual growth rate of MFP can be calculated in the usual way:

$$\ln\left(\frac{A_i(t)}{A_i(t-1)}\right) = \ln\left(\frac{Y_i(t)}{Y_i(t-1)}\right) - (1-\bar{\beta})\ln\left(\frac{C_i(t)}{C_i(t-1)}\right) - \bar{\beta}\ln\left(\frac{L_i(t)}{L_i(t-1)}\right) \quad \text{(A.4)}$$

Notes

1 Without pretending to be exhaustive, of the noteworthy contributions one may look at Nadiri (1980), Griliches (1984), Cuneo and Mairesse (1984), Patel and Soete (1988), Verspagen (1995a) and Higón (2007).

2 See Englander and Mittelstädt (1988), Lichtenberg (1992), Coe and Helpman (1995) and Ulku (2007).

3 This chapter based on Antonelli *et al.* (2010).

4 Hence knowledge fungibility and knowledge compositeness can be considered two aspects of knowledge recombination (Antonelli, 2008; Antonelli and Calderini, 2008; Antonelli *et al.*, 2008).

5 The limitations of patent statistics as indicators of innovation activities are well known and include their sector-specificity, existence of non-patentable innovations and the fact that there are other protection tools. Moreover, the propensity to patent varies over time as a function of patenting cost, and is more likely to feature large firms (Pavitt, 1985; Levin *et al.*, 1987; Griliches, 1990). Nevertheless, patents can be useful measures of new knowledge production especially in the context of analyses of aggregate innovation performance (Acs *et al.*, 2002). There is also debate over patents being considered an output rather than an input of innovation activity and empirical analysis shows that patents and R&D are dominated by a contemporaneous relationship, further supporting use of patents as a proxy for innovation (Hall *et al.*, 1986). Patent application is a time- and resource-consuming process, likely to produce ex-ante selection of the innovations to be patented which enables identification of high-value innovations stemming from systematic and more formalized innovation efforts, which are the object of our analysis.

6 Since Jaffe (1986, 1989), technological fields have been used to calculate technology-related variables. Out of the former empirical studies using IPC codes assigned to European Patents it is worth recalling the one by Verspagen (1997). More recently IPC codes have been successfully employed in empirical analyses to calculate technological variety and relatedness (see Breschi *et al.*, 2003; Nesta and Saviotti, 2005, 2006; Nesta, 2008).

7 This depreciation rate is very common in empirical analyses that derives the knowledge stock either from R&D investments (Griliches, 1990; Loos and Verspagen, 2000) or from patent applications (Nesta, 2008).

8 It is worth noting that the inclusion of the RTA index in econometric specifications may yield some biased estimates (Laursen, 1998). This is due to the fact that the index squeezes the values signalling non-specialization between 0 and 1, while values signalling specialization are between 1 and infinity. This gives rise to a skewed distribution that in turn implies the violation of the normality assumptions of the error term in the regression. For this reason it is recommended to use some transformation of the index that makes its distribution close to the normal one. In the following econometric estimations we have taken standardized values for the RTA, the distribution of which proximate very much normality.

9 We are aware this may introduce a "home bias" in the analysis, which could be solved by considering triadic patents. Unfortunately, we are not able to extract the same set of information about triadic patents and thus are obliged to limit our analysis to European patents.

10 The most common means of assigning patents to territorial units is by inventor's address. Following this procedure is important when analysing the effects of knowledge spillovers on innovation performance. Although our dataset is quite detailed, we do not have information on inventors' addresses. However, we analyse the effects of changes to the internal structure of the knowledge base on productivity growth and, therefore, on the use of technological knowledge. Thus, we do not expect that this problem significantly affects our estimates.

11 We acknowledge that these may turn out to be very strong assumptions. Nonetheless, such approach, fairly common in the literature about the determinants of productivity growth, has the advantage of allowing for the calculation of different inputs' elasticities for each country at each time. It therefore accounts for cross-sectional and time variation.

7 Evolutionary patterns of knowledge structure in biotechnology

7.1 Introduction

The economic systems of advanced capitalist societies have been facing a gradual process of transition towards the so-called knowledge-based economy. In this context the creation and utilization of knowledge become the key factors affecting the competitiveness of firms, regions and countries (Freeman and Soete, 1997). The novelty of the knowledge-based economy has often been exaggerated in two senses: (1) knowledge has always been used in every production process and in every human activity; (2) however restricted may be the meaning that we wish to attach to knowledge, the so-called knowledge-based economy started to be developed in the second half of the nineteenth century. What differentiates a modern knowledge-based economy from a more traditional one is the process by which the knowledge used in production processes is generated. Starting from the mid nineteenth century, institutions specialized in the production and diffusion of knowledge were created. Examples of these institutions are the German, or von Humboldt, university system and industrial research and development. The process of knowledge production in these institutions differed from the more traditional one in which knowledge had always been created as a joint product of activities having a different objective, such as the production of a material output. A modern knowledge-based economy is thus characterized by the growing percentage of knowledge used in production processes which comes from specialized institutions.[1]

In view of this, the study of the mechanisms of knowledge production has received renewed attention in the last decade, while considerable effort is today dedicated to characterize the knowledge base of different sectors in the economy and to detect its impact on firm performance and on industrial organization (Breschi *et al.*, 2003; Krafft, 2004; Nesta and Saviotti, 2005; Corrocher *et al.*, 2007).

The chapter develops an approach to technological knowledge as a collective good within an evolutionary framework. Knowledge is characterized by a high degree of cumulativeness and recombination across a number of different subunits. Knowledge is thus viewed as a retrieval–interpretative and a co-relational structure. The sectors' knowledge base, as well as their internal structure, may be represented as networks the nodes of which are the technological classes cited in patent documents, while the link between two nodes represents the co-occurrence of the technological classes in the same patent. Some recent theories of knowledge have stressed the recombinant aspect of knowledge (Fleming and Sorenson, 2001; Olsson, 2000). Although we share the view that a very large proportion of discoveries and innovations can be created by the recombination of existing ideas, we think that not all the examples of genuine novelty can be originated by recombining existing knowledge. The network-based representation of knowledge we propose in this chapter can accommodate both views: genuine novelty will be represented by the emergence of completely new nodes while the recombination of existing ideas can be represented by the creation of new links between existing nodes

This chapter aims at applying the methodological tools of SNA to the analysis of the structure of knowledge bases and of their evolution over time, so as to identify their usefulness in the identification of the emergence of discontinuities in the technology lifecycles.[2] We will also see that SNA tools capture the transition of technological activities from exploration to exploitation strategies characterized by organized search rather than random screening. To this purpose we will compare the results obtained by applying SNA with those of previous research.

We will focus on the dynamics of knowledge bases within one specific knowledge-intensive sector along the 1980s and 1990s, i.e. biotechnology, and on its relationship with its lifecycle. The analysis is conducted by using information contained in patent applications submitted to the European Patent Office (EPO), which are drawn by the Espacenet database. In doing this we do not neglect the essential distinction between knowledge and information. At a conceptual level, patents are more likely to be considered as representations of the inventors' knowledge than *stricto sensu* knowledge themselves. However, we follow previous empirical studies that highlight the usefulness of patents as measures of production of new knowledge, beside R&D intensity, or other measures of innovation output (Acs *et al.*, 2002; Griliches, 1990). We use the EPO data to map the frequency of co-occurrences of technological classes within patents and to calculate a number of indexes, i.e. information entropy, knowledge coherence and cognitive distance, on the one hand, and network density, degree, closeness and betweenness, on the other.

Our results show the existence of interesting and meaningful similarities between the two sets of empirical indicators with reference both to the relative levels of the variables and their evolution over time. Such similarities allow us to link the evolution of SNA measures to the different phases of lifecycles the industry underwent in the period of observation.

The rest of the chapter is organized as follows. In Section 7.2 we elaborate on the concept of knowledge networks and spell out the working hypotheses. Section 7.3 describes the data and the methodology. Section 7.4 introduces the measures of social network analysis (SNA). In Section 7.5 and 7.6 we provide the empirical results of our analysis. Finally, in Section 7.7 we discuss the results and provide the conclusions.

7.2 Knowledge networks

From the discussion conducted in Chapter 4 we can deduce that knowledge can be represented as a network the nodes of which are variables, connected by links determined by the joint utilization of different variables. New knowledge stems from the creative recombination of heterogeneous bits of knowledge, which are fragmented and dispersed among economic agents. The set of the elements (nodes) and their interactions (links) making up the network can be defined as the structure of the knowledge base, which is characterized by a specific architecture at each point in time. This is in turn both an effect and a determinant of the interactions among agents involved in the collective process of knowledge creation (Krafft and Quatraro, 2011). Agents are hardly aware of the architecture of the knowledge network, and there is by no means any agent able to command all the relevant knowledge on the topology of the technology space. The combination of the different bits of knowledge yields results, i.e. the architecture of the network, which "if they were brought about deliberately, would require a knowledge on the part of the directing mind which no single man can possess" (Hayek, 1937, p. 52).

A network has a number of properties. For example, the emergence of new concepts and variables, leading to the creation of new nodes, is likely to affect network density unless the rate of creation of new links is equal to the rate of creation of new nodes. In general we can expect these two rates to differ systematically during the different phases of the lifecycle of a given knowledge type (Saviotti, 2009). Knowledge establishes connections between variables, provided these variables exist. Thus, we expect the creation of new nodes to precede the creation of at least a part of their link: new and still poorly connected nodes will emerge during the early phases of a discontinuity and the rate of creation of links will pick up later during the normalization or maturation phase. Network density could then be expected to fall at the emergence of a discontinuity and to rise during the subsequent maturation of knowledge. If we consider the network of knowledge in its entirety, given the above-described dynamics of the creation of nodes and of links, we can never expect it to be completely connected. New variables are likely to be created in different regions of knowledge space, corresponding to different disciplines, before all the possible connections are established. In other words, the rates of creation of new nodes in the network of knowledge cannot be expected to coincide at all times with the rate of creation of links. As a consequence network density becomes a relevant variable to characterize the dynamics of knowledge.

The structure of knowledge network may therefore be regarded as an emergent property stemming from qualified interactions among innovating agents. As such, its architecture is likely to change in the course of time. We can expect the evolution of the network of knowledge to occur in a number of ways, from an initial discontinuity to the recomposition of a new network: (1) new concepts and variables, which will be represented as new nodes, emerge; (2) some old concepts and variables become extinct; (3) new connections are established between new or old concepts and variables, giving rise to corresponding new links; (4) the relative weight of old and new nodes and links changes in the course of time.

The possibility to represent knowledge as a network provides an adequate conceptual foundation for the study of processes of knowledge generation and utilization in firms and industries. To identify all the variables and the connections present in the knowledge base of a firm at the lowest possible level of aggregation would be a prohibitively expensive task. An approximate version can then consist of identifying relatively "small" units of knowledge and their connections. We identify these "small" units within the traces of knowledge which have been used so far, such as patents and publications.

In this chapter we adopt a more explicit network approach to the representation of the KB. We consider knowledge as an integrated system, in which both the constituting elements and the connections among them deserve to be investigated. The representation of the KB as a network enables us to better appreciate the dynamics of the emergence of new knowledge types by monitoring the changes in nodes and links. If we allow the nodes to represent technological classes and the links to represent the interactions of technological classes within the same patent, the dynamics of network density provides useful evidence about the relationship between the growth of technological classes and the growth of the corresponding links.

As we said above, it is reasonable to expect the increase in technological classes not to be followed immediately by a proportionate increase in the links among them. This leads us to expect network density to fall over time when the growth rate of the variety of technological classes is higher than the growth rate of the variety of connections. The rate of creation of nodes at the onset of a discontinuity can be expected to be higher than the rate of creation of links. However, this trend cannot persist indefinitely. New nodes cannot continue to be isolated or poorly connected since the production of artefacts requires the joint utilization of several types of knowledge, which are then by definition complementary. The full exploitation of the new knowledge types requires an increase in the number of links per node. This increase can be expected when the rate of creation of new nodes slows down, possibly even to zero, but we cannot exclude that it can happen even if the number of technological classes, and thus of new nodes, keeps increasing.

In fact, this is a situation that could not be easily based on simple dichotomies such as exploration/exploitation or random/organized search, but that our quantitative approach to the properties of knowledge allows us to articulate better. New nodes can represent types of knowledge radically or slightly different from the existing ones. The latter situation would occur, for example, when the new types of knowledge are obtained by specialization of pre-existing ones with which they would share the basic concepts. These two situations can be distinguished by their cognitive distance: the emergence of radically different nodes would correspond to a high cognitive distance while that of slightly different nodes would correspond to a low cognitive distance. We can also expect the construction of links between radically different nodes to require a greater effort and a longer time than the construction of links between slightly different nodes. In this respect the distinction between related and unrelated variety is extremely useful: when related variety dominates we can expect the number of links to grow at a rate comparable to or even higher than that of the number of nodes while the number of links would always grow at a lower rate than the number of nodes when unrelated variety dominates. Different types of nodes can generate different cognitive distances depending on whether they are slightly or radically different from pre-existing nodes.

In addition to network density, the toolbox of SNA contains also interesting measures to characterize the relative weight of nodes, and hence of technological classes, and the related changes over time. Such measures are referred to as "centrality measures". Out of these, the *degree*, the *closeness* and the *betweenness* are the most commonly used. The concept of centrality refers to the relative importance, or weight, of a node within a network. Different measures of centrality are available depending on whether one wishes to measure it at the *local* or at the *global* level within the network. Degree centrality is the most local of these measures as it is based on the relative number of links of a node with its neighbours. Closeness builds upon the geodesic distance of a node from all the other nodes in the network. Should a node be directly connected with every other node, its closeness centrality would be very high. It is straightforward that high average levels of closeness are likely to correspond to high average levels of degree. Betweenness measures the relative importance of a node over the whole network. It builds upon a triadic relationship, according to which a node is central as long as it represents a kind of unavoidable stop in the paths connecting any other pair of nodes in the network.

Empirical observations of the knowledge base of firms show that at any time the distribution of nodes around links is very uneven (Saviotti, 2009). Some types of knowledge are relatively more important than others. There is no a priori reason to expect sectoral knowledge bases to behave differently. When a discontinuity emerges we can expect a fall in network density but the evolution of the structure of the network is more difficult to predict. Some old nodes, including important ones, are going to disappear and new nodes are going to emerge, some of which will become important. We have already described this as an example of structural change in knowledge. However, it is more difficult to say whether the number of important nodes is going to rise or to fall since it depends among other things on variety of the knowledge base. When more new nodes emerge than old ones disappear the number of important nodes is likely to grow. We can see this problem as the analogue of industrial concentration: in most cases the distribution of the centrality of nodes will resemble an oligopoly, with few nodes having many links and being very central and with the majority of nodes having a low centrality. The evolution of both centrality and of average centrality measures is difficult to predict since it depends on the combination of a number of factors including the growth in the number of nodes, the growth in the number of links, the rate of growth of variety, the ratio related/unrelated variety, cognitive distance, etc.

In this chapter we map and measure the KB of sectors rather than of firms. In this case the KB we map depends on inter-individual and inter-organizational interactions both at the intra- and at the inter-firm level. Since the sector is a population of broadly comparable firms to have a complete representation of it we would need to measure both the means and the distribution of the properties of the KB within the population. For reasons of space in the present chapter we describe only the patterns of evolution reflecting the behaviour of the average or representative firm.

On the basis of the previous considerations we can now formulate the following three propositions:

P1: The emergence of a discontinuity in a type of knowledge suitable to become the future knowledge base of a sector leads to the sequence of the two periods of random search first occurring in the exploration phase, and of organized search later in the exploitation phase.

P2: During the random search period we expect overall knowledge variety to rise and to be dominated by unrelated variety, coherence to fall and cognitive distance to rise. As the maturation of the new technology subsequently begins we expect variety to keep rising or falling but to be dominated by related variety, coherence to rise and cognitive distance to fall.

P3: At the onset of a knowledge discontinuity we expect the rate of creation of new nodes to exceed the rate of creation of new links and the density of the network of knowledge to fall. As the maturation of the new technology subsequently begins we expect the rate of creation of new links to start exceeding the rate of creation of new nodes and the density of the network of knowledge to start rising (Saviotti, 2009).

For the time being, it is very difficult to make any predictions about the time path of the various centrality measures or about the evolution of the structure of knowledge. We will come back to this point in the discussion of our results.

Before concluding this section let us remark that a knowledge discontinuity has very important implications for the management of a firm which uses this knowledge. The more dissimilar the new knowledge is with respect to the firm's previous KB, the lower the absorptive capacity of the firm for it will be with its present human resources. In order to internalize the new knowledge the firm would need to hire completely new human resources familiar with the new knowledge and probably to lay off a large part of its existing human resources which have now become redundant. Needless to say, this is neither an easy operation nor one which can be carried out at great speed. Furthermore, the larger the incumbent firm, the more difficult this transformation of its knowledge base and of its human resources is likely to be. This would at least partly explain the emergence of dedicated biotechnology firms (DBFs) and the formation of innovation networks with large diversified firms.

7.3 Data and methodology

7.3.1 Measurement of the knowledge base

The information concerning patent applications required to test the working hypotheses formulated in Section 7.2 was obtained from the Espacenet database provided by the European Patent Office.[3] The initial dataset consisted of 2,659,301 items, including both EU and worldwide applications, over the period 1978–2005. The analysis thus focuses on the subset of patent applications concerning the biotechnology sector, which has been identified by merging the classifications set up by the OECD and by the French *Observatoire des Sciences et des Techniques*. We adopted these classifications to establish some tentative boundaries for the biotechnology sector, although we acknowledge that in some cases these classifications leave some important classes out.

Our search strategy is based on queries reporting the IPC classes that define biotechnology. Taking into account these elements, it resulted that the sector includes 11 IPC classes, reported in Table 7.1.[4]

The total number of patent applications in the biotechnology sector amounts to 321,449. Figure 7.1 represents the dynamics of patent applications, by considering

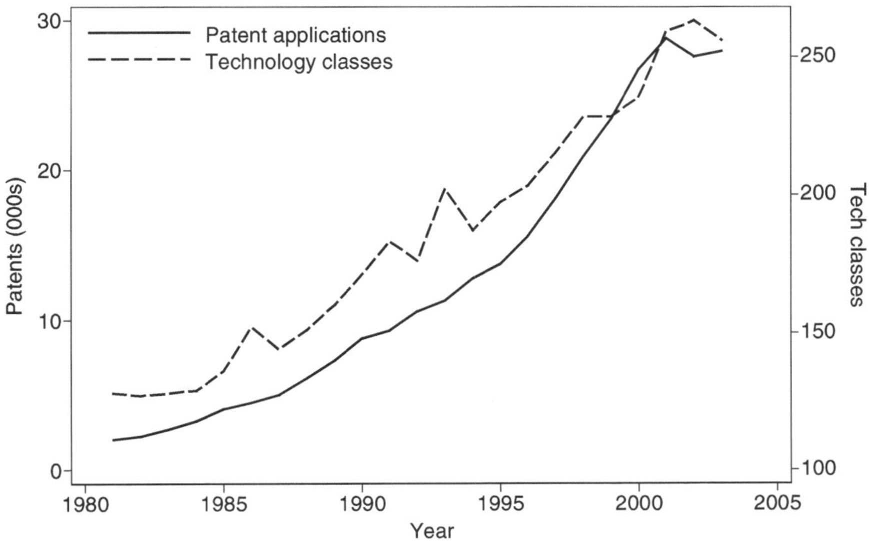

Figure 7.1 Dynamics of patent applications and technological classes in biotechnology.

Table 7.1 Definition of the biotechnology sector using IPC classes

A01H	New plants or processes for obtaining them; plant reproduction by tissue culture techniques
A61K	Preparations for medical, dental or toilet purposes
C02F	Treatment of water, waste water, sewage or sludge
C07G	Compounds of unknown constitution
C07K	Peptides
C12M	Apparatus for enzymology or microbiology
C12N	Micro-organisms or enzymes; compositions thereof
C12P	Fermentation or enzyme-using processes to synthesise a desired chemical compound or composition or to separate optical isomers from a racemic mixture
C12Q	Measuring or testing processes involving enzymes or micro-organisms; compositions or test papers thererof; processes of preparing such compositions; condition-responsive control in microbiological or enzymological processes
C12S	Processes using enzymes or micro-organisms to liberate, separate or purify a pre-existing compound or; processes using enzymes or micro-organisms to treat textiles or to clean solid surfaces of materials
G01N	Investigating or analysing materials by determining their chemical or physical properties

the five-year cumulated number, and the related number of observed technological classes. It is clear that the number of patent applications (on the left y-axis) increased over the entire period at an increasing rate, showing no discontinuities in the series. The evidence concerning technological classes is slightly different (on the right y-axis). The rate of growth indeed appears to be slower than in the case of patent applications. Moreover, the pattern of evolution over time presents almost regular discontinuities in 1986, 1991, 1994 and 1999.

The number of technological classes may be considered an approximate measure of diversity. It is to be observed that the informational entropy function which we used to measure technological variety measures in fact a combination of variety and balance since it is affected by both the total number of classes and by the extent of their diffusion. On the other hand, the informational entropy function cannot take into account disparity (Stirling, 2007). Disparity is the most difficult component of diversity to measure since it refers to the extent of intrinsic difference between two entities. One could argue that to measure disparity is impossible since it would amount to provide a quantitative estimate of qualitative change. In general we would expect radical innovations to have a greater disparity than incremental innovations. However, we do not have a criterion to compare the disparity of two different radical innovations. The distinction between related and unrelated variety helps us in this respect since it defines two sets knowledge with different disparity, higher for unrelated variety and lower for related variety.

The slower rate of change of the number of technological classes relative to that of patents can be interpreted as a sign of the growing maturation of biotechnological knowledge. This finding corresponds well to the declining rate of growth of technological variety occurring in the second half of the 1990s. The observed discontinuities are likely to be linked to changes in the internal structure of the knowledge base. Changes of this type occurred during our period of observation.

In the rest of the chapter we combine two different approaches to study the knowledge base of the biotechnology sector. On the one hand, we take into account the results obtained by measuring properties of knowledge such as variety, coherence and cognitive distance, which draw upon co-occurrence matrices. The measurement of these properties was implicitly based on knowledge being represented as a network but it did not explicitly use SNA. On the other hand, we explicitly mobilize SNA in the field of economics of knowledge. The emphasis of this method rests on the architecture of networks and on the characterization of each node with respect to the other ones. Plenty of applications can be found in the economic literature, above all in the study of interactions among different kind of agents within industrial and technological districts (Morrison, 2008; Giuliani, 2007). To our best knowledge, there are no attempts to apply this methodology to the investigation of the recombinant dynamics underlying knowledge generation and utilization.[5]

In this context, we can think of nodes as technological classes, whereby a link between two nodes represent the co-occurrence of technological classes within the same patent. The network of relationships among the nodes provides an image of the internal structure of the knowledge base of the sector under scrutiny, i.e. biotechnology. Given a dataset of patent applications, one may represent the evolution of the knowledge base by deriving a network for each observed year, and calculating the relevant indexes accordingly. This allows us to characterize technological classes according to their relative position in the structure of the knowledge base, and to investigate the pattern of change over time. We will propose an interpretation of the main concepts and indicators typical of SNA presented in Section 6.3 in terms of knowledge-related dynamics.

The usefulness of SNA for the investigation of the dynamics of knowledge bases can be better appreciated by directly comparing the two approaches, and emphasizing differences and similarities between the two sets of indicators. We accomplish this task in Section 7.4, in which we present the results of our calculations.

7.4 Empirical results

7.4.1 Using co-occurrences matrices

In this section we will develop an analysis of the knowledge of biotechnology according to the measures described in Section 6.2. The first aspect that we want to investigate in the results of our calculations is the presence of a transition from random to organized search. To test the existence of this transition we constructed a co-occurrence matrix of the technologies used in the patents awarded to the three knowledge intensive sectors in our database. Each patent is classified according to a primary and to a number of secondary classes. Such matrices are constructed by assigning frequencies to the couples of IPC classes occurring together. If the transition from random to organized search occurs, we expect a declining fraction of the off diagonal cells to contain a growing share of the overall frequency of co-occurring technologies. In other words, the transition from random to organized search should involve a process of concentration of the technological choices made in the patents. In a graphic representation of the co-occurrence matrix (Figure 7.2) this phenomenon is revealed by a growing share of few and higher peaks among those representing all the possible technological combinations.

While the comparison among the four diagrams of Figure 7.2 reveals the interesting evidence of an increasing concentration, it needs to be complemented by an analysis of the characteristics of knowledge structure in order to better grasp the lifecycle dynamics of the sector. We can immediately notice that the technological variety of biotechnology rises during the period 1981–2003 (Figure 7.3a). Unrelated variety dominates between 1981 and 1983 and related variety becomes dominant between 1983 and 2003. Moreover, the rate of growth of variety falls for most of the period of observation until it becomes constant from the early 1990s, with the possible exception of the mid-1980s. In 1985 the rate of growth of variety starts rising in correspondence with the overtaking of unrelated variety by related variety. In our case while in the early 1980s the unrelated variety was higher than the related, the situation was reversed starting from 1985. This would suggest that, while in the very early phases of the emergence of modern biotechnology most of the new knowledge was coming from outside the knowledge base previously used, starting from 1985 internal (to the sector) sources of knowledge differentiation became more prominent. However, it must be observed that starting from the mid-1990s a trend began to the convergence of related and unrelated variety. This trend is likely to be caused by the emergence of a second generation of biotechnology linked to bioinformatics, a new type of competence coming from a discipline different from biology.

Coherence starts with a very low value in 1981 and rises, although with some fluctuations, during the whole period of observation (Figure 7.3b). In this case as well as in all the other measures of properties of the knowledge base we can distinguish within the overall changes a trend and superimposed deviations. The deviations are probably due to a combination of real events affecting the dynamics of knowledge and of noise due to the quality of the data. Thus, we cannot expect all the deviations to be easily interpretable. Both variety and coherence show an overall positive trend accompanied by superimposed deviations. In particular, there are two periods of fast rise in knowledge coherence, beginning in 1982 and in 1995 respectively. The first of these deviations from the trend seems to be closely related to the ratio of related to unrelated variety. When unrelated variety is greater than the related one, in the period 1981–1982, the coherence index falls. It then begins to increase in 1983 when related variety overtakes unrelated variety. The subsequent rise in 1997 cannot be explained in the same way. However, it can be observed that the two rises in knowledge coherence seem to coincide with the onset of the absorption of two different generations of biotechnology, based on recombinant DNA and on genomics respectively, by incumbent firms (Saviotti and Catherine, 2008). The transition between the two generations led to a discontinuity in the pattern of inter-firm alliances: within each generation the number of alliances followed a lifecycle, increasing first, reaching a maximum and then declining. The competences required in the two generations differed as bioinformatics acquired a key role in the sequencing of genomes.

Taking this into account we can interpret the overall rising trend in knowledge coherence as due to the growing relative similarity, or low cognitive distance, of the new types of knowledge which incumbent firms needed to learn. The deviations with respect to the trend could be explained by the emergence of new generations of biotechnology and/or by the ratio of intra- to inter-group variety. As a new generation of biotechnology emerges the overall trend is not reversed but deviations can occur due to the however limited cognitive distance that the new generation introduces. This line of explanation is not incompatible with the one based on the ratio of related to unrelated variety. We can assume changes in related variety to involve a more limited change in coherence than those in unrelated variety because the former can be obtained by recombination and differentiation of the same concepts while the latter are more likely to involve the introduction of completely new concepts. In other words, a rise in related variety is likely to involve a lower extent of knowledge discontinuity than an equivalent rise in unrelated variety and to lead to lower fall in coherence. Conversely we can expect changes of generation within one technology (e.g. biotechnology) to raise the ratio related/unrelated while the emergence of a completely new technology can be expected to lower the same ratio. However, in

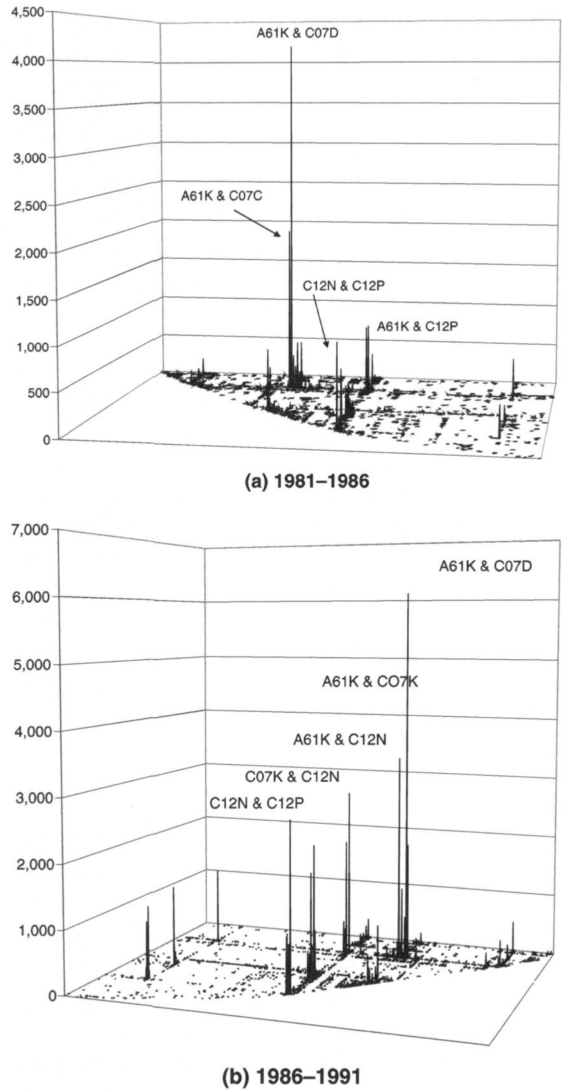

Figure 7.2 Matrix of co-occurrences in the biotechnology sector.

some cases the situation can be more complex. In this context the transition between the two generations of biotechnology involved two contrasting trends: the second generation shared the same basic biological concepts with the first generation but required the use of competences and concepts in bioinformatics which were new to biologists and which came from another discipline. We can expect the first trend to raise both related variety and coherence and the second to reduce both of them. What we observe is then the result of a trade-off between

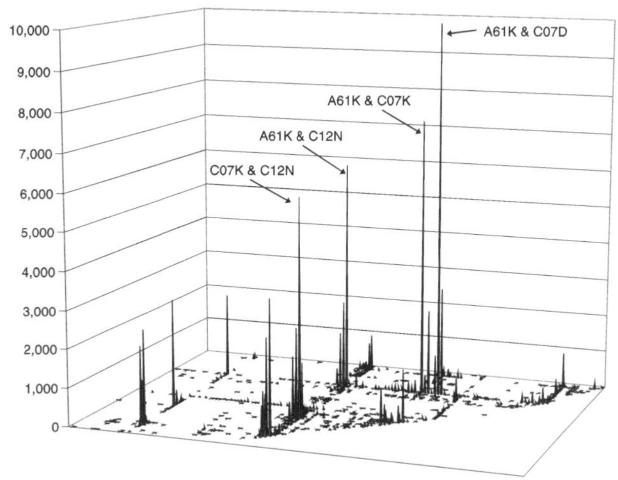

(c) 1991–1996

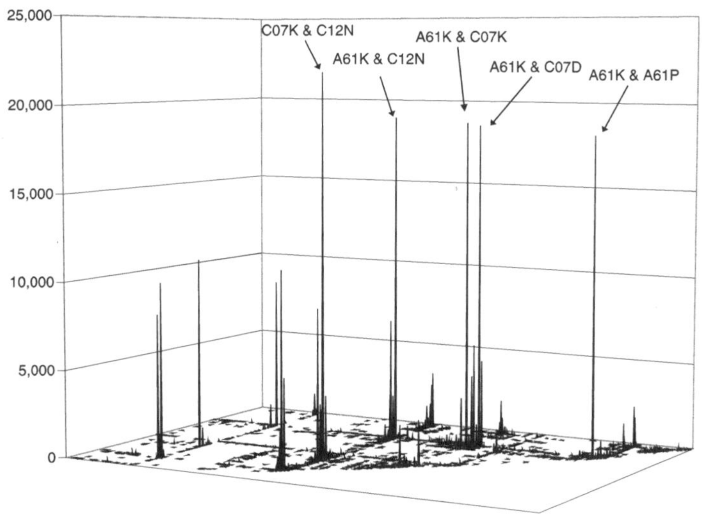

(d) 1996–2001

Figure 7.2 continued

the two trends described above. This interpretation is compatible with (1) the tendency to the convergence of related and unrelated variety beginning in the mid-1990s and (2) the slowdown in the rate of growth of coherence between 1988 and 1996 followed by a rise in coherence beginning in 1997, which could be due to the maturation of the second generation of biotechnology.

Cognitive distance falls during the whole period of observation (Figure 7.3c). These results can be interpreted as the consequence of the knowledge discontinuity which occurred in the early 1970s with the emergence of what is called

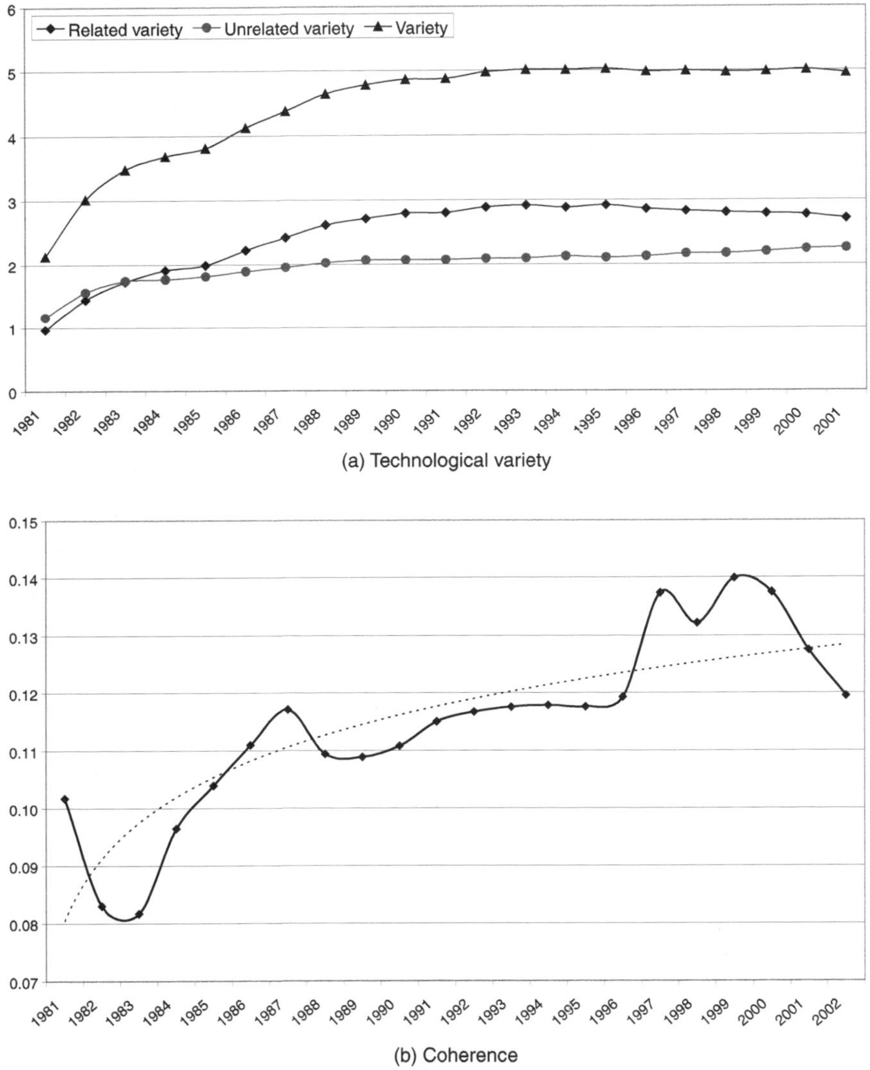

Figure 7.3 Properties of knowledge base of biotechnology.

third-generation biotechnology, linked mostly to the first industrial applications of molecular biology. We expect this knowledge discontinuity (1) to have raised the technological variety of biotechnology using firms (then mostly pharmaceutical and agrochemical) by adding to their KBs new technological classes, (2) to have initially reduced the coherence of the same firms since the new technological classes were initially poorly connected to the pre-existing ones, (3) to have initially raised the cognitive distance by adding new technological classes which were very dissimilar from those previously used by incumbent firms. We expect these phenomena to have occurred immediately after the onset of the knowledge discontinuity, a period corresponding to random search or to exploration, but for which unfortunately we have no data. The evolution that we can trace in Figure 7.3 corresponds to the beginning of the maturation of biotechnological knowledge. The process of diversification of the KB proceeds but it shifts away from the more radical innovations corresponding to unrelated variety to the more incremental and local ones corresponding to related variety. Contrary to what would have occurred if variety had remained mostly unrelated, coherence can now start rising and cognitive distance can now start falling as the process of knowledge diversification occurs by the more incremental and local changes corresponding to related variety. These findings confirm that the emergence of a knowledge discontinuity starts a lifecycle in which initially unrelated variety and cognitive distance rise and coherence falls. In the subsequent part of the lifecycle unrelated variety rises to become dominant, coherence rises and cognitive distance falls.

It is important to point out that without the distinction between related and unrelated knowledge variety the simultaneous occurrence of rising overall

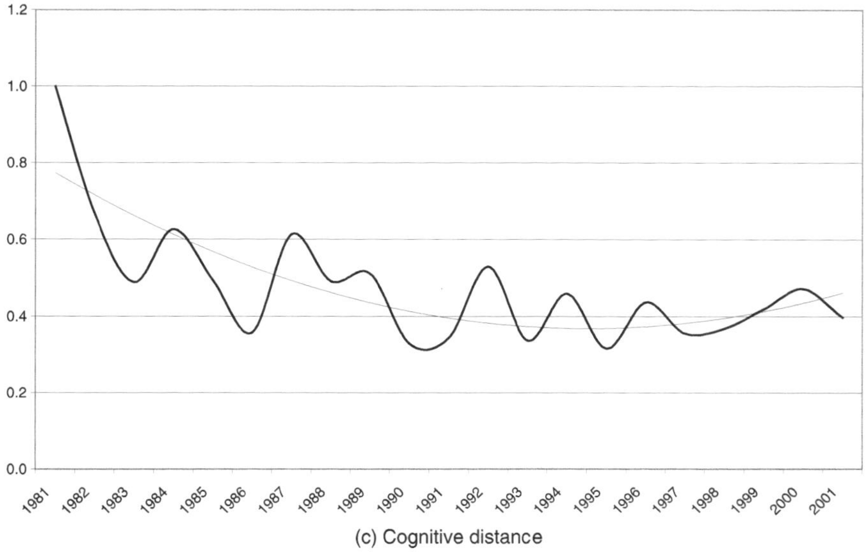

Figure 7.3 continued.

variety, rising coherence and falling cognitive distance would have been very difficult to explain. The distinction between related and unrelated variety turns out to be as fruitful in the study of structural change in knowledge as it is in the study of structural change in economic systems (see Frenken *et al.*, 2007; Saviotti and Frenken, 2008). This is a further example of the greater subtlety that we can achieve by means of our measures of properties of knowledge.

7.4.2 The implementation of SNA: networks and knowledge structure

In order to calculate the density and the centrality indexes described in Section 5.3 we have rearranged the dataset so as to make it suitable for processing by means of Pajek software. After having chosen patent life to last for five years, we have split the dataset in order to obtain a network for each observed year, the nodes of which are technological classes and links represent the co-occurrence of technological classes within the same patent documents. Since we are investigating the relationships occurring among "actors" belonging to the same set or class, we have derived "mode one" networks. Moreover, it must be noted that, in a given year, two technology classes may occur together in more than one patent application. This would imply the presence of multiple links between two nodes. While this represents useful information, the calculation of density and centrality measures requires multiple lines to be removed, so as to obtain unbiased results. However, the graphical analysis presented in the following section will help appreciating multiple links as a proxy of the strength of relationships among nodes, by making the thickness of edges proportional to observed frequency of technology couples.

Let us start analysing the structure of knowledge base by looking at the dynamics of network density, which is reported in Figure 7.4. The range of variation of the index is between 0.045 and 0.064, while the average is about 0.054. Density falls from 1983 to 1991 and then it starts growing until 2001. However, these two periods are not characterized by a smooth dynamics. On the contrary, a number of discontinuities can be observed, both in the decreasing and in the increasing periods. Let us first concentrate on the main trend and then try to explain the discontinuities. We can notice in Figure 7.3a that the knowledge property which shows the best correlation with density is total technological variety which rises between 1981 and 1991 and remains constant afterwards. Thus, density falls when technological variety rises and starts rising when technological variety becomes constant. The main trend of density in the period studied corresponds to our predictions concerning the rates of growth of the number of nodes and of the number of links. We expect the number of nodes to grow faster than the number of links immediately after the discontinuity and the number of links to start growing faster as the new type of knowledge moves towards maturity. The inversion from negative to positive of the slope of the density curve occurs when the rate of growth of total technological variety becomes zero and when the number of technological classes per patent starts declining. In this case the relative rates of growth of related and of unrelated variety do not seem to be the main factor determining the evolution of density. At best the ratio related variety/unrelated variety (RTV/UTV) could have determined the early discontinuity occurring in 1986, when RTV first overtook UTV, and the later slow down in the rate of growth of density occurring in 1992 when the ratio RTV/UTV started declining.

Let us now proceed to analyse the results obtained with the SNA approach. Here we have two types of information: first, we have somewhat more aggregate measures of centrality, such as degree, closeness and betweenness; second, we have a finer representation of the structural change occurring in knowledge by means of the network of technological classes at different times. In the latter we can see the emergence of new technological classes, the decline or extinction of older ones, the change in the pattern of links and the consequent change in the distribution of links around nodes. These measures and representations help us interpreting the evolution of biotechnology knowledge. Furthermore, they need to be related to the above-mentioned properties of knowledge which they should extend and complement.

We start by describing the pattern of change in centrality measures. To this purpose we have first calculated degree centrality, betweenness and closeness for all technological classes. Then we sorted them at each year according to the observed values for each of the indexes. Finally, at each year we kept only the top ten classes, for each variable. In so doing, we built three matrices that are reported in Table 7.2, Table 7.3 and Table 7.4, which can be read both horizontally and vertically. In the columns one can appreciate the dynamics of

Figure 7.4 Dynamics of network density for biotechnology.

technological classes over time, identifying whether they have been central all over the period or only in some years. By looking at the rows one can appreciate the change in the structure of knowledge base, with respect to the composition of the group of most central technological changes.

Let us look at the data concerning the normalized degree centrality in Table 7.2. According to this index, the classes showing the highest degree are A61K and C02F. While the former is a market-oriented class, the latter is related to environment-friendly technologies for the treatment of waste water. The dynamic evidence for the two classes is very similar and characterized by a limited fluctuation over time. The case of the C12N class is interesting in that its degree centrality has increased of about the 70 per cent over the whole time period, and can therefore be described as the technology characterized by the best dynamics. This class involves the study of micro-organisms, carrier-bound enzymes and genetic engineering. Thus the gradual rise of its centrality, above all in the 1990s, is the signal of the increasing recombination of such class of technologies with the rest of the technologies that make up the structure of the knowledge base.

From the systemic viewpoint, one may note that there are six classes that appear in our top ten at each observed year. These may be defined as the core of the knowledge base, within which we have noted genetic engineering has gained increased relevance with respect to more established technologies. Moreover, we have some classes that mainly appear in the first decade, like A23L and C07C, and some classes that mainly occur in the 1990s, like C07K, C12Q and G01N. The first two refer to the treatment and preservation of food and to organic chemistry compounds (mainly hydrocarbons). The second group refers to peptides, to the composition and the preparation of testing processes involving enzymes or micro-organisms, and to physics testing technologies useful to investigate the micro-structure of materials.

Table 7.3 reports the data concerning the closeness centrality. This measure is the inverse of geodesic distance, and it may be thought of as the average distance of a node from all the others. The maximum value of closeness for a node is reached when it is directly connected with the rest of the network. Thus, it seems reasonable to expect that the degree and the closeness of a node are strictly related each other. Indeed the picture is almost the same as in the previous table. The classes showing the highest closeness are again A61K and C02F, though their dynamics is characterized by limited fluctuations. The only class showing a clear-cut increasing trend over time is the C12N, the closeness of which grows about 8 per cent in 20 years (thus this evidence is less pronounced than in the case of degree).

The systemic layout also resembles the one provided by the degree index. Still one can note the persistence of the six classes described before as the core of the structure of the knowledge base. The same also applies to the pattern of emergence and disappearance of classes over time. This evidence thus supports the idea that the structure of the knowledge base of the biotechnology sector has been characterized by the existence of a strong core, a sort of building block,

Table 7.2 Dynamics of normalized degree centrality, top ten technological classes

	A01N	*A23L*	*A61K*	*A61L*	*B01D*	*B01J*	*C02F*	*C07C*	*C07K*	*C08F*	*C12M*	*C12N*	*C12P*	*C12Q*	*G01N*
1981	0.2262	0.2143	0.6071		0.3690	0.2679	0.6310	0.2619		0.2202		0.2917	0.3095		
1982	0.2742	0.2473	0.6559		0.3871	0.3011	0.6129	0.2688		0.2312		0.3280	0.3065		
1983	0.3109	0.2642	0.6269		0.3886	0.3161	0.6425	0.2798				0.3523	0.2902		0.2280
1984	0.3100	0.2500	0.6250		0.4050	0.3350	0.6250	0.2600				0.3750	0.2950		0.2300
1985	0.3112	0.2407	0.6266		0.4315	0.3444	0.6473	0.2448				0.3817	0.2988		0.2282
1986	0.2353	0.2127	0.5837		0.3575	0.2760	0.6290		0.1991			0.3620	0.2805		0.2081
1987	0.2353	0.2036	0.5882		0.3982	0.2805	0.6290		0.2217			0.3575	0.2941		0.2172
1988	0.2912	0.2386	0.6351	0.2526	0.4246	0.3368	0.6386	0.2421				0.3754	0.2947		
1989	0.2195	0.2398	0.6057	0.2114	0.3293	0.2276	0.5894		0.2398			0.3293	0.2846		
1990	0.2567	0.2375	0.6092		0.3257	0.2605	0.5632		0.2452			0.3410	0.2720	0.2337	
1991	0.2500	0.2500	0.5993		0.3272	0.2316	0.5588		0.2868			0.3640	0.2831	0.2610	
1992	0.2688		0.5914		0.3548	0.2616	0.6057		0.3082			0.3907	0.2975	0.2903	0.2796
1993	0.2508		0.5974		0.3432	0.2574	0.5875		0.3036			0.3861	0.2805	0.2838	0.2937
1994	0.2601		0.5912		0.3514	0.2804	0.6081		0.3142			0.4189	0.2973	0.3041	0.3311
1995	0.2630		0.6021		0.3322	0.3080	0.6159		0.2976			0.4187	0.3080	0.3287	0.3460
1996	0.2413		0.5874		0.3322	0.3287	0.6119		0.3182			0.3986	0.3182	0.3427	0.3601
1997	0.2690	0.2552	0.6000		0.4103	0.3310	0.6552		0.3034			0.3828	0.2690		0.2759
1998			0.5860		0.3509	0.3193	0.6456		0.3053		0.2561	0.3895	0.3018	0.3474	0.3509
1999		0.2570	0.6021		0.3697	0.3134	0.6585		0.3099			0.3873	0.2993	0.3099	0.3345
2000	0.2757		0.6176		0.3750	0.3162	0.6507		0.2978		0.2868	0.4265	0.2757		0.2978
2001	0.2799		0.6231		0.3619	0.3097	0.6157		0.3022			0.4216	0.2687	0.2649	0.2910

Table 7.3 Dynamics of closeness centrality, top ten technological classes

	A01N	*A23L*	*A61K*	*A61L*	*B01D*	*B01J*	*C02F*	*C07C*	*C07K*	*C08F*	*C12M*	*C12N*	*C12P*	*C12Q*	*G01N*
1981	0.5600	0.5581	0.7179		0.6131	0.5773	0.7304	0.5753		0.5619		0.5854	0.5915		
1982	0.5759	0.5688	0.7440		0.6200	0.5886	0.7209	0.5776		0.5653		0.5981	0.5905		
1983	0.5920	0.5744	0.7283		0.6206	0.5938	0.7366	0.5796				0.6069	0.5848		0.5643
1984	0.5917	0.5682	0.7273		0.6270	0.6006	0.7273	0.5747				0.6154	0.5865		0.5650
1985	0.5921	0.5657	0.7281		0.6376	0.6040	0.7370	0.5697				0.6179	0.5878		0.5644
1986	0.5667	0.5539	0.7038		0.6071	0.5785	0.7246		0.5553			0.6105	0.5816		0.5567
1987	0.5667	0.5525	0.7083		0.6225	0.5801	0.7246		0.5623			0.6088	0.5862		0.5595
1988	0.5852	0.5666	0.7326	0.5711	0.6333	0.6000	0.7308	0.5677				0.6156	0.5864		
1989	0.5616	0.5668	0.7172		0.5971	0.5629	0.7069		0.5681			0.5985	0.5829		0.5591
1990	0.5736	0.5649	0.7190		0.5945	0.5724	0.6941				0.5637	0.6028	0.5787	0.5662	
1991	0.5702	0.5714	0.7139		0.5939	0.5608	0.6869					0.6112	0.5824	0.5751	0.5631
1992	0.5753		0.7099		0.6078	0.5741	0.7136		0.5911			0.6214	0.5874	0.5849	0.5813
1993	0.5695		0.7129		0.6036	0.5728	0.7047		0.5895			0.6196	0.5816	0.5827	0.5861
1994	0.5748		0.7098		0.6066	0.5804	0.7167		0.5932			0.6325	0.5873	0.5896	0.5992
1995	0.5757		0.7153		0.5996	0.5898	0.7225		0.5874			0.6324	0.5910	0.5983	0.6046
1996	0.5686		0.7079		0.5996	0.5971	0.7204		0.5946			0.6245	0.5946	0.6034	0.6098
1997	0.5777		0.7143		0.6291	0.5992	0.7417		0.5894			0.6183	0.5777	0.5720	0.5800
1998			0.7072		0.6064	0.5938	0.7383		0.5901		0.5711	0.6209	0.5888	0.6051	0.6064
1999		0.5703	0.7154		0.6134	0.5917	0.7454		0.5917			0.6201	0.5880	0.5917	0.6004
2000	0.5787		0.7234		0.6154	0.5926	0.7411		0.5875		0.5824	0.6355	0.5800		0.5862
2001	0.5801		0.7263		0.6091	0.5903	0.7204		0.5890			0.6336	0.5776	0.5763	0.5852

which is constituted by the most important classes of the period. While the existence of a core confirms the uneven distribution of technological classes in the knowledge base of the sector the composition of the core changes in the course of time with some older classes becoming extinct or losing importance and with some new ones emerging and becoming important components of the knowledge network. Classes linked to food preservation and to organic chemistry are examples of the former, classes linked to molecular biology or to physical measurements are examples of the latter.

Table 7.4 shows indeed the results for the calculation of betweenness centrality. The emerging picture is slightly different in this case, as compared to closeness and degree. Now one can distinguish one dominating class, i.e. the C02F, the dynamics of which is pretty stable over time. The A61K class, although showing high values, is characterized by a decreasing trend over time. Two results deserve special attention. First, the betweenness centrality of the C12N class grows by 150 per cent in 20 years. This means that this class has become more and more relevant not only with respect to its direct links to other classes, but also as a "gatekeeper" that allows for indirect recombination among technologies within the knowledge base.

Second, the systemic properties of the knowledge structure are differently characterized by this index. Indeed one may note that the core classes are now seven instead of six. The additional core class is the G01N, which is the physics class related to the investigation of micro-materials. Moreover the A23L is no longer listed in the top ten of central classes, while C12M (related to the investigation of enzymes and micro-organisms) appears already in 1983 and remains until 2001. We can then conclude that betweenness emphasizes more the global influence of technological classes over the network of knowledge while degree and closeness focus more on their local influence.

The analysis of the dynamics of centrality measures characterizing technological classes has revealed two important aspects. First of all, even in a period of pronounced structural change and of knowledge discontinuities the knowledge base of biotechnology is characterized by an apparently stable structure, in which one may identify a limited number of core technologies, around which there is a dynamics of emerging and disappearing classes. However, it is worth stressing that changes in the relative centrality of technological classes occur also within the core itself. Thus, the structure of the core is affected by qualitative change over time. Moreover, the closeness and the degree centrality show very similar patterns, while the betweenness centrality seem to provide a somewhat different evidence in terms of dynamics of centrality and of systemic features. Thus, change occurs but it is not instantaneous. This reflects the difficulty and costs inherent in transforming the knowledge base of whole sectors.

Such aspects may be better grasped by looking at the average centrality measures, calculated according to Equation (5.6). The three diagrams of Figure 7.5 report the dynamics of weighted average degree, closeness and betweenness respectively, and the distribution of technological classes (represented by the scattered points) around the average values (i.e. the continuous lines). Even in

Table 7.4 Dynamics of betweenness centrality, top ten technological classes

	A01N	*A61K*	*A61L*	*B01D*	*B01J*	*C02F*	*C07C*	*C07K*	*C08F*	*C09K*	*C12M*	*C12N*	*C12P*	*C12Q*	*G01N*
1981		0.3179		0.0838	0.0369	0.3961	0.0187		0.0190	0.0192		0.0410	0.0535		0.0261
1982	0.0243	0.3222	0.0285	0.0889	0.0513	0.3211	0.0202					0.0528	0.0343		0.0254
1983	0.0314	0.2701	0.0275	0.0779	0.0478	0.3638	0.0225					0.0587	0.0273		0.0262
1984	0.0301	0.2848	0.0196	0.0862	0.0478	0.3343					0.0290	0.0743	0.0325		0.0197
1985	0.0226	0.2777	0.0183	0.0915	0.0463	0.3404					0.0212	0.0654	0.0280		0.0253
1986	0.0191	0.3013		0.0736	0.0334	0.3984		0.0184			0.0415	0.0696	0.0331		0.0161
1987	0.0168	0.3053		0.0849	0.0282	0.3921		0.0201			0.0333	0.0667	0.0306		0.0164
1988		0.2931	0.0216	0.0870	0.0405	0.3169					0.0232	0.0679	0.0263	0.0200	0.0245
1989		0.3117		0.0708	0.0253	0.3532		0.0295			0.0371	0.0753	0.0328	0.0240	0.0276
1990	0.0238	0.3114		0.0650	0.0336	0.3382					0.0521	0.0779	0.0307	0.0434	0.0234
1991		0.2892		0.0673	0.0239	0.3257		0.0403			0.0598	0.0801	0.0308	0.0518	0.0285
1992		0.2502		0.0712	0.0287	0.3287		0.0525			0.0481	0.0815	0.0271	0.0506	0.0304
1993		0.2954		0.0707	0.0258	0.3274		0.0495			0.0466	0.0751	0.0248	0.0457	0.0379
1994		0.2761		0.0679	0.0290	0.3205		0.0485			0.0419	0.0798	0.0269	0.0439	0.0456
1995		0.2664		0.0506	0.0364	0.3245		0.0350			0.0417	0.0761	0.0328	0.0442	0.0478
1996		0.2420		0.0513	0.0431	0.3485		0.0357			0.0432	0.0649	0.0388	0.0517	0.0512
1997		0.2522		0.0748	0.0397	0.3323		0.0395			0.0378	0.0606	0.0195	0.0274	0.0364
1998		0.2246		0.0561	0.0358	0.3646		0.0238			0.0563	0.0551	0.0259	0.0542	0.0568
1999		0.2312		0.0616	0.0301	0.3720		0.0301			0.0429	0.0504	0.0256	0.0446	0.0528
2000		0.2413		0.0659	0.0294	0.3502		0.0337			0.0462	0.0868	0.0165	0.0260	0.0327
2001		0.2605		0.0683	0.0398	0.3155		0.0457			0.0399	0.1004	0.0184	0.0202	0.0297

this case it seems clear that degree and closeness are characterized by very similar patterns. An evident cyclical fluctuation may indeed be noted in the first decade in both cases, followed by a relatively more stable dynamics in the second half of the 1990s. The dynamics of average betweenness is instead characterize by definitely less pronounced fluctuations, and by a decreasing trend over the whole period.

An important point worth noting is that the centrality measures have a bimodal distribution, shown by the separate sets of points at the top and bottom of Figure 7.5. The part of the distribution at the bottom of the figure contains a very large number of points while the part at the top contains fewer and more scattered points. This confirms the extreme skewedness of the distribution of links around nodes and seems to correspond to the description of this distribution as an oligopoly with few highly connected and many poorly connected technological classes. It is to be noticed also that the distribution is even more skewed for betweenness than for degree and closeness. Thus, an even smaller proportion of technological classes is globally, as opposed to locally, important in the network of knowledge.

7.5 Graphical analysis of networks: the web of knowledge

In addition to the more aggregate measures of density and of centrality SNA allows us to explore the fine structure of knowledge and the changes it undergoes in the course of time by showing the changes in the types and weight of nodes and of links. This is the most direct way in which we can estimate the extent of structural change which is occurring in the knowledge base of the biotechnology sector.

The first half of the 1980s is characterized by a relatively simple network structure. It can be easily seen in Figure 7.6 that the core node in the network corresponds to the class A61K, i.e. to a very generic and market-oriented class referring to medical preparations and cosmetics. The two important links of A61K are those with the classes C07C "Organic Chemistry" (1,576 co-occurrences) and C07D "Heterocyclic compounds" (3,236 co-occurrences). Also the direct arc connecting these two classes shows a pretty high frequency (573). Although at smaller magnitudes, other relevant nodes are C12P, C12N and C12R, which are combined both with each other and with A61K. This structure reflects the nature of the knowledge base which was predominantly used at the beginning of our period of observation. It is to be pointed out that although DBFs have played an extremely important role in the development of biotechnology, and especially in the early period (see Grabowski and Vernon, 1994), the KB we detect is likely to be affected much more by the KBs of the large incumbent firm which have many more patents. Thus, it is natural for the sectoral KB of the early 1980s to contain mostly classes related to organic chemistry or to market-related classes which are known to have constituted the KB of large pharmaceutical and agrochemical firms before that time.

In the second half of the 1980s the network takes a slightly more complex form, due to the emergence of additional nodes. The connection between A61K and C07D is still the most recurrent, as it is observed 6,096 times. The co-occurrences of A61K with C12P (fermentation and synthesis of compounds) and C12N (micro-organisms and enzymes) gain momentum in this period, the latter moving towards the third rank. Also, in this period the G01N class becomes a more important node in the network, well connected with the other relevant

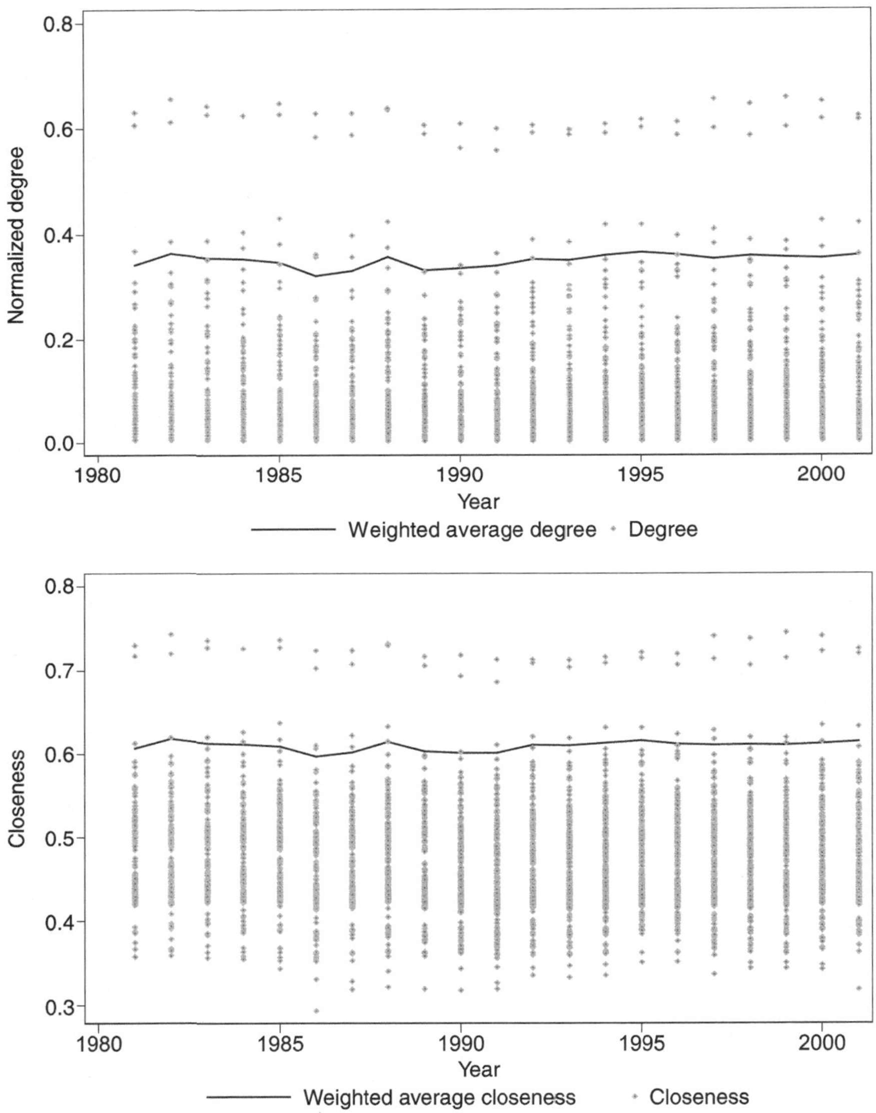

Figure 7.5 Average centrality measures.

nodes, in particular with the A61K and C12N classes. It must be noted that a new class emerges as relevant, i.e. C07K (peptides). This shows a very high degree of connectivity with A61K, so much that this couple is now the second most recurrent in the network. This emerging class is also well connected to C12P and C12N, so that now we might say that the core of the biotechnology activity is characterized by a set of four, or at least five classes, which are directly or indirectly connected to all other classes in the network.

Figure 7.6 shows the network of the period 1991–1995. The network structure appears now to be consolidated, in the footsteps of the configuration the sector reached in the previous period. In particular, the weight of the C07K class further increases, so that it can be considered as a persistent hub, besides C12N, C12P, C07D and most of all A61K. In this picture also the G01N preserves its position, as a class that is neither marginal nor very central. It would seem to play a supporting role for all other classes. Two new relevant classes deserve to be mentioned here, i.e. C07H (nucleosides) and C12Q (measuring or testing processes involving enzymes or micro-organisms).

In the last period we observe, i.e. the second half of the 1990s, the network would seem to be slightly more complex. We still observe one single class which acts as "core" class, i.e. the A61K. Then we may observe a set of second level classes, which have a central position although not as central as the A61K. Such classes are C12N, C07K and A61P. Then there is a third level, made up of nodes which still show a good degree of connectivity, but are slightly peripheral, like the C12P, C12Q, C07H and G01N. One could say that while in the first period

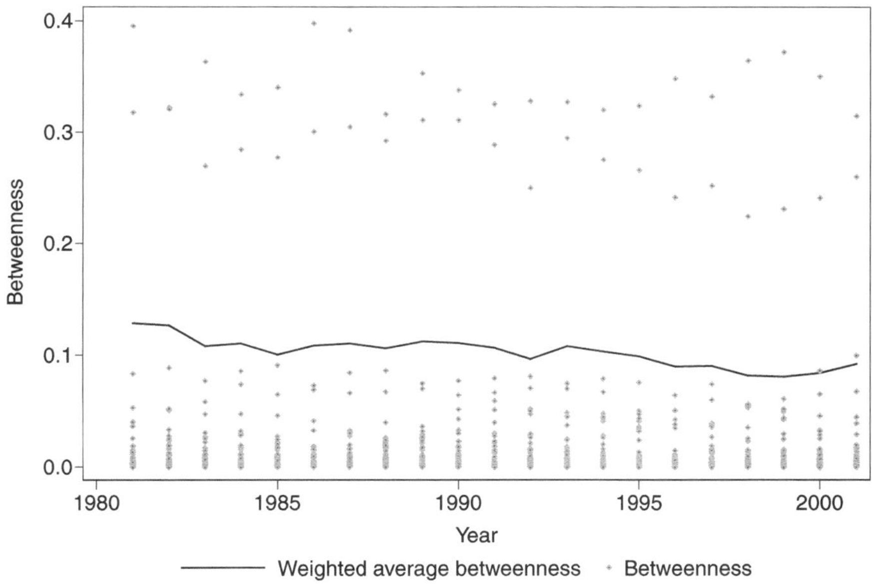

Figure 7.5 continued.

the network showed a very high level of concentration, it has become more distributed over time, but characterized by a kind of hierarchical structure.

In summary, during the period 1981–2000 the network of biotechnological knowledge undergoes a structural change in which some technological classes linked to the previous knowledge base of pharmaceutical and agrochemical firms, at that time the main users of biotechnology, disappear or lose importance

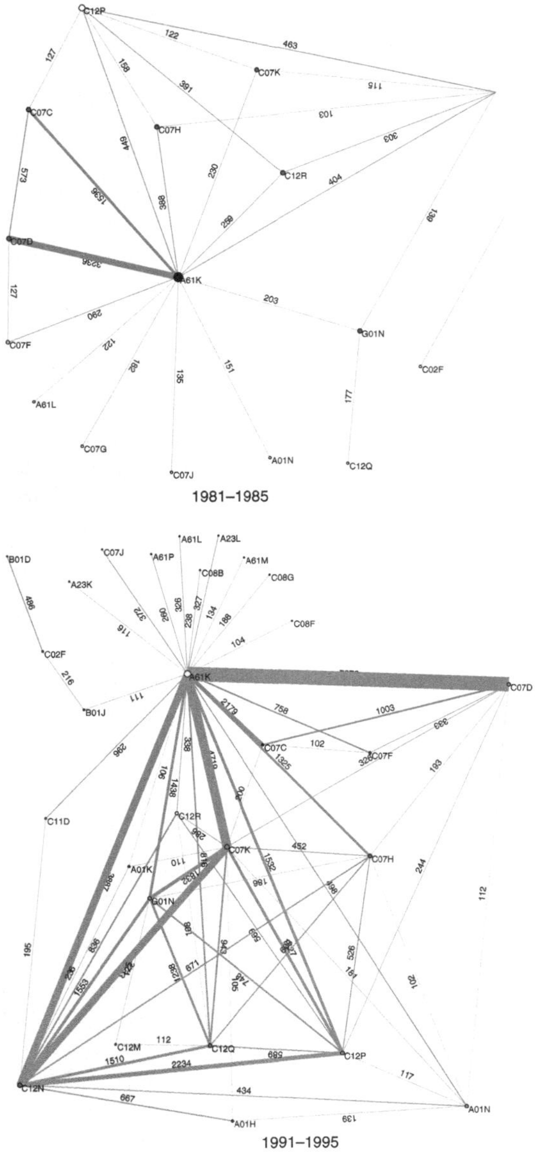

Figure 7.6 Network of technology classes for biotechnology, four sub-periods.

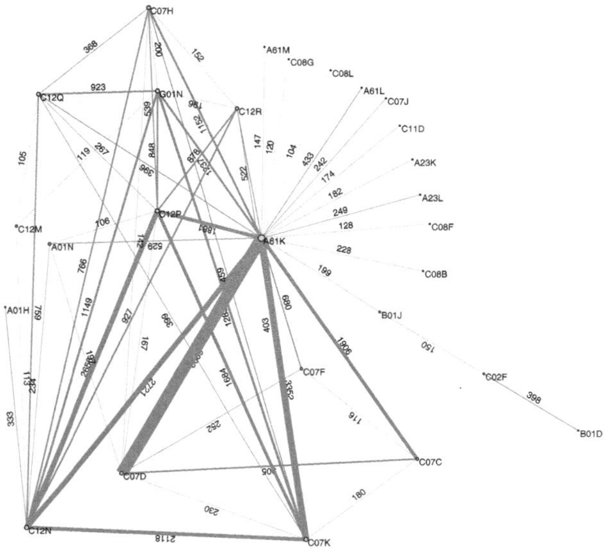

1986–1990

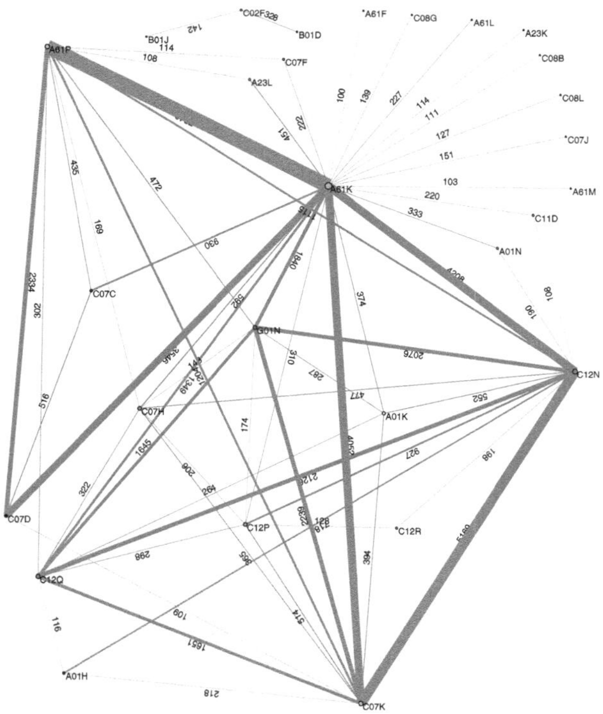

1996–2000

Figure 7.6 continued.

and other classes emerge and acquire a greater weight in the network. The older and declining classes corresponded mostly to organic chemistry, which until the 1970s constituted the KB of pharmaceutical and agrochemical firms. The newer and emerging classes correspond to molecular biology and to physical measurements, which have become the core of the new biotechnology. Three points are worth noting here: first, the process of structural transformation of the KB has been fairly slow; second, although many of the classes corresponding to the old KB have disappeared, some remain and are still of considerable importance (see CO7C and C07D); third, the knowledge network of biotechnology has a hierarchical structure with a very skewed distribution of links around nodes. However, the network seems to have become more polycentric in the course of time, with a growing number of relatively important nodes. This is likely to be due to the growing number of technological classes.

7.6 Discussion and conclusions

In this chapter we studied the dynamics of knowledge generation in biotechnology. We mapped the knowledge base of this sector by means of the patents awarded by the EPO during the period 1981–2002. We did not distinguish the different types of economic actors to which the patents were given but considered the sector as a whole. We have characterized the structure of the knowledge base by drawing upon SNA. Our analysis included the measure of four network properties, density, degree, closeness and betweenness, and the graphic representation of the network of knowledge at different times during the period 1981–2000. We combined this analysis based on SNA with the results of previous research in technological variety, related and unrelated, the coherence and the cognitive distance of biotechnological knowledge were measured using the same set of data.

We interpreted our results as showing that the knowledge base of biotechnology using firms, mostly pharmaceutical and agrochemical, was affected in the 1970s by a discontinuity constituted by the discovery of recombinant DNA and monoclonal antibodies, which suddenly shortened the time horizon during which industrial applications could be expected. This discovery event had required a very long period of preparation in which the research leading to the creation of a new discipline (molecular biology) began, in the 1930s, and in the end led to the critical events which catalysed the first industrial applications. In order to adequately study the evolution of knowledge in biotechnology our data would have needed to cover most of the 1970s. Given the limitations of our data for the time being we have to infer what is likely to have happened before the beginning of our period of observation. In biotechnology, based on the very low initial value of both variety and coherence and on the fact that coherence was still falling at the beginning of the period of observation, we expect unrelated variety to have been greater than related variety during all of the 1970s and until 1983. Thus, the 1970s would have been the period when the discontinuity in biotechnological knowledge constituted by the adoption of molecular biology would have first manifested itself and the 1980s the period during which the new knowledge started to be adequately integrated into the knowledge base of biotechnology using firms. In the early 1970s incumbent pharmaceutical and agrochemical firms found themselves faced with the very difficult task of learning a new type of knowledge for which they had a very low absorptive capacity. As a consequence the internalization of the new knowledge was slow and gradual but eventually it led to the extinction or decline of some old technological classes and to the incorporation of some new ones. From our results it appears that biotechnology progressively enters into a more mature phase of development.

The emergence and subsequent impact of a knowledge discontinuity creates a lifecycle beginning with the birth of the discontinuity and ending once the new knowledge has become a routinized component of the KB. This lifecycle can be described by a number of concepts, such as random or organized search, exploration or exploitation, revolutionary or normal science. These concepts are highly suggestive and very helpful in organizing our thoughts but they are not analytically rigorous. The properties of the knowledge base that we measure in our chapter provide a means to make these concepts more analytical. Thus, we expect to be able to explain the transition from exploration to exploitation based on our measurable properties. In fact, since the previous transition can correspond to more than one time pattern of the properties we measure, concepts like exploration or exploitation can provide a broad brush-stroke representation of a process into which our quantitative approach allows us to detect much finer details. Thus, we could say that biotechnology has already entered a more mature phase in which exploitation related activities tend to grow with respect to exploration related ones. During this phase the rate of growth of technological variety gradually falls, related variety overtakes unrelated variety, coherence rises and cognitive distance falls. We expect these trends to correspond either to organized search or to exploitation since a fast rise of overall variety is dominated by the unrelated type, while a fall in coherence and a rise in cognitive distance are also observed. However, we cannot be certain about the exact correspondence of the above trends in knowledge properties and the phases of the lifecycle. Past work showed us that different combinations or trends of knowledge properties can correspond to each of the concepts exploration, exploitation, random or organized search.

The transition to the organized search period seems to occur as some particularly fruitful research trajectories emerge, which are then followed by the majority of participants. The evidence about the established properties confirms that the biotechnology sector has undergone such a transition in the past 20 years (Krafft *et al.*, 2009). Moreover, and most importantly here, the approach based on SNA proved to be a very useful means to investigate the changing structure of the KB. Network density turned out to fall between 1981 and 1991 and to rise afterwards until 2000. This result corresponds closely to our expectations according to which network density should fall in the early phases of a discontinuity when the rate of growth of new technological classes, and therefore of new nodes, is expected to be higher than the rate of growth of new links. Network density can be expected to start rising when the new knowledge starts maturing and the rate of growth of links overtakes the rate of growth of nodes. Various measures of centrality confirm the results previously obtained with properties such as variety, coherence and cognitive distance.

The technological classes which turned out to be important in the previous study occupy the most central positions in the network of knowledge and their evolution corresponds closely to our previous observations. However, the graphic representation of networks of knowledge and the various centrality measures that SNA allows us to greatly enhance our ability to detect patterns. For example, we find that the market-oriented A61K class retains a very high local centrality during the whole period while its betweenness starts falling. Thus, the A61K class remains very central but it loses its ability to act as a gate-keeper over the whole network. Also, the calculation of average centrality measures shows that the distribution of the various centrality measures for the different classes is clearly bimodal, a finding which fits very nicely with the observation that few technological classes have many links and most technological classes have very few links. This provides further support to the idea that there cannot be devised any a priori assumption on the relationships between the locus of the filière to which a class may be assigned and the observed centrality. On the contrary, one could well observe a shift of high centrality values from downstream to upstream classes in relationship to the actual stage of technology lifecycles, providing further information on the dynamics of knowledge-intensive sectors.

With this chapter we have extended previous attempts to explore the dynamics of knowledge in a knowledge intensive sector like biotechnology. Here we have added to the measures of the knowledge properties previously developed (variety, coherence, cognitive distance) an approach based on SNA. This new approach confirms and extends our previous results. For example, by means of SNA we can measure changes in network density and distinguish between different measures of centrality, which we could not do with our previous toolbox.

The methods we describe and the results we obtain seem to us very important to develop the tools required to represent and measure knowledge as we move towards a knowledge-based economy and society. It is important to stress that the representation of the network of knowledge we used in this chapter has both some specific features and some general features common to other systems. Namely, if the structure of the system is defined by its elements (nodes) and by their interactions (links), then the emergence of a set of completely new concepts gives rise to a *discontinuity* in the evolution of knowledge, and further to the emergence of a new paradigm or a new research programme based on completely novel ideas. Indeed, a discontinuity can be expected to have on the overall time profile of knowledge an effect similar to the emergence of a paradigm. In fact, we can say that the revolutionary phase of a paradigm results from the emergence of a discontinuity (Kuhn, 1962). Within the paradigm the revolutionary phase would be followed by a period of normal science, during which a more incremental pattern of knowledge accumulation would occur (Kuhn, 1962).

Of course, we realize that this is very preliminary work and that, although our findings suggest some general conclusions, they will need to be further tested and better articulated.

Notes

1 In spite of this clearly established trend, knowledge produced according to more traditional methods is still used alongside the one produced in specialized institutions. Thus, the operational definition of knowledge that we propose in this chapter cannot establish a demarcation between scientific and non-scientific knowledge but needs to be generally applicable to all the types of knowledge which can be combined in human activities.

2 This chapter is based on Krafft *et al.* (2011).

3 We consider thus patent applications as the best indicator of firms' knowledge bases, though the usual caveats mentioned in the literature may apply. We use these data to map the frequency of co-occurrences of technological classes within patents and to calculate a number of indexes, i.e. information entropy used to measure related and unrelated variety, knowledge coherence and cognitive distance.

4 Though the use of IPC classes to define sectors' boundaries may present some drawbacks, as they are function-oriented (Corrocher *et al.*, 2007), the merging of two classifications allows our study to be much more inclusive than many other studies, and reduces the risk of neglecting important classes. It is worth noting that these classes include quite different technologies and processes, which might be placed at different stages of an ideal filière of the knowledge production process. This is a potential source of misunderstandings or misinterpretation of our results, due to the fact that one could claim that classes in certain stages of such filière are more likely to be central than classes impinging upon other stages. However, given the interactive nature of the knowledge-creation process, this may help more the discussion of empirical results than the ex-ante formulation of expectations.

5 It is fair to note that a similar approach has been attempted at the firm level by Yayavaram and Ahuja (2008).

8 Knowledge, structural change and productivity

A special focus on Italian regions

8.1 Introduction

Since the seminal contributions by Nelson (1959) and Arrow (1962), knowledge has attracted more and more the attention of economists, both with respect to the mechanisms leading to its production, dissemination and exchange, and with respect to its effects on productivity.

Despite this, empirical contributions estimating the effects of knowledge on economic growth has appeared only after the path-breaking works by Zvi Griliches (1979). Within this strand of literature, the traditional production function has been extended so as to include knowledge as an additional input. Knowledge is conceived as a bundled stock, as if it were the outcome of a quite homogeneous and fluid process of accumulation made possible by R&D investments, the same way as capital stock.[1]

Empirical analyses at the regional level have instead appeared quite recently. These mainly focus on the determinants of cross-regional differences in the efficiency of knowledge creation, like knowledge spillovers and spatial proximity, within the context of a knowledge production function approach (Acs *et al.*, 2002; Fritsch, 2002, 2004; Fritsch and Franke, 2004; Crescenzi *et al.*, 2007).

Yet, to the best of the author's knowledge, no empirical investigations can be found in literature analysing the effects of technological knowledge on regional growth.

This chapter aims at bringing technological knowledge into an empirical framework analysing the determinants of cross-regional differential growth rates.[2] To this purpose, we consider technological knowledge as the outcome of a combinatorial search activity carried out across a technological space in which combinable elements reside (Weitzman, 1998; Fleming, 2001; Fleming and Sorenson, 2001). In this direction we are able to specify a set of properties that can describe the internal structure of the regional knowledge base and that go beyond the traditional measure of knowledge capital stock. Indicators like knowledge coherence and knowledge variety can be calculated by exploiting the information contained in patent documents, and in particular by looking at the co-occurrence of technological classes which patents are assigned to (Saviotti, 2007). While studies can be found investigating these properties at the firm level (Nesta and Saviotti, 2006; Nesta, 2008), and at the sectoral level (Krafft *et al.*, 2009; Antonelli *et al.*, 2010), there is no empirical evidence at the regional level yet.

Our analysis focuses on the effects of knowledge dynamics on the evolution of the manufacturing sector within Italian regions over the period 1981–2002.[3] This appears to be a particularly appropriate context for our purposes. Indeed, the Italian economic structure has long been characterized by a sharp dualism. On the one hand North-West regions were the cradle of modern industrial firms, and during the 1980s the manufacturing sectors had already completed their growth phase, leaving the floor to service industries. On the other hand, North-Eastern-Central (NEC) regions showed a delayed development of manufacturing activities, carried out mostly by small and medium-sized enterprises (SMEs) often operating in peculiar economic and social environments (Fuà, 1983). The role of innovation on such cross-regional differences have become the object of empirical analysis only recently (Quatraro, 2009a, 2009b), and the investigation of knowledge dynamics in this framework may provide useful insights to gain a better understanding.

In this context, the contribution of this chapter to the literature is threefold. First, it applies to notion of recombinant knowledge at the regional level, by identifying a set of properties able to define the structure of the architecture of regional knowledge bases. Second, such analysis is relevant for its general implications concerning the relationships between the dynamics of technological knowledge and regional growth, in particular with respect to regional innovation strategies. Finally, it also aims at rejuvenating a field of inquiry which has been lacking appropriate consideration since the 1980s. For this reason, the debate about the economic development of Italian regions has missed the important opportunity of investigating cross-regional differences in the light of the economics of knowledge and innovation.

The rest of the chapter is organized as follows. In Section 8.2 we outline the theoretical framework and propose a model linking regional productivity growth to the characteristics of knowledge base. Section 8.3 presents the methodology and Section 8.4 describes the regional knowledge indicators. In Section 8.5 we describe the data sources and provide descriptive statistics for the main variables. Section 8.6 presents the results of the empirical estimations and an extension to spatial panel data models, while Section 8.7 provides a discussion of results in the light of the Italian economic history. Finally, conclusions and policy implications follow in Section 8.8.

8.2 The model

The discussion articulated in Chapter 4 dynamics of technological knowledge can therefore be understood as the patterns of change in its own internal structure, i.e. in the patterns of recombination across the elements in the knowledge space. This allows for qualifying both the cumulative character of knowledge creation and the key role played by the properties describing knowledge structure, as well as for linking them to the relative stage of development of a technological trajectory

(Dosi, 1982; Saviotti, 2004, 2007; Krafft *et al.*, 2009). Moreover, the grafting of this approach into the analysis of the determinants of cross-regional growth differentials allows for a better understanding of the interplay of knowledge dynamics and the patterns of regional industrial development. The ability to engage in a search process within cognitive spaces that are distant from the original starting point is likely to generate breakthroughs stemming from the combination of brand new components (Nightingale, 1998; Fleming, 2001; Fleming and Sorenson, 2001; Sorenson *et al.*, 2006). In this direction regional innovation capabilities may be defined as the ability of regional actors to engage in the combinatorial process that gives rise to the structure of the regional knowledge base (Lawson and Lorenz, 1999; Romijn and Albu, 2002; Antonelli, 2008).

The economic development of regions is indeed strictly related to the innovative potentials of the industries they are specialized in. Firms within a propulsive industry grow at faster rates, propagating the positive effects across firms directly and indirectly related to the propulsive industry. The potentials for creating new knowledge are at the basis of regional growth, and they happen to be unevenly distributed across sectors according to the relative stage of lifecycle (Perroux, 1955; Kuznets, 1930; Burns, 1934; Schumpeter, 1939).[4]

The intertwining of industrial and technological lifecycles is therefore of great importance, as well as the distinction between exploration and exploitation (March, 1991). The introduction of new technologies is indeed more likely to show a boosting effect on economic performances as long as the search activity enters an exploitation stage wherein potential dominant designs are selected and implemented. The creation of new knowledge in this phase, and hence the resulting knowledge base, is more likely to involve by the recombination of knowledge bits characterized by a great deal of complementarity and by the identification of diverse and yet highly related knowledge bits. A further dichotomy between random screening and organized search seems to be relevant in this direction. The transition to organized search is typical of phases in which profitable technological trajectories have been identified, and the recombination activity occurs out of a sharply defined region of the knowledge space. The likelihood of successful innovations is greater in this stage, and marks the difference between mature and growing sectors (Krafft *et al.*, 2009, 2011).

The discussion conducted above leads us to propose a simple model to appreciate the effects of the properties of knowledge structure on regional economic growth:

$$g_{i,t} = f(K_{i,t-1}) \tag{8.1}$$

Where subscripts i and t refer respectively to the region and to time, g is the growth rate of productivity and K is the regional knowledge base. Traditionally, K is defined as the stock of knowledge corrected for technical obsolescence: $K_{i,t}=\dot{k}_{i,t}+(1-\delta)K_{i,t-1}$, where $\dot{k}_{i,t}$ is the flow of new knowledge at time t and δ is the rate of obsolescence. This relationship is able to capture the influence only of intangible capital, neglecting the characteristics of regional knowledge.

In order to appreciate the implications of the recombinant knowledge approach on the operationalization of the properties of knowledge structure, the K term of Equation (8.1) can be modelled by extending to the regional domain the framework that Nesta (2008) develops at firm level. Let us recall the main passages in what follows.

Assume that a region is a bundle of D productive activities, represented by the vector $P=[p_1,\ldots, p_d,\ldots, p_D]$. Each regional activity p_d draws mainly upon a core scientific and technological expertise e_d, so that the regional total expertise is the vector $E=[e_1,\ldots, e_d,\ldots, e_D]$. The regional knowledge base emerges out of a local search process aimed at combining different and yet related technologies. This implies that an activity p_d may also take advantage of the expertise developed in other activities $l(1\neq d)$, depending on the level of relatedness τ between the technical expertise e_d and e_l. It follows that the knowledge base k used by the dth activity is:

$$k_d \equiv e_d + \sum_{l\neq d}^{D} e_l \tau_{ld} \tag{8.2}$$

The meaning of Equation (8.2) is straightforward. The knowledge base k of each activity d amounts to the sum of its own expertise and the expertise developed by other activities weighted by their associate relatedness. Such equation can be generalized at the regional level to define the aggregate knowledge base:

$$K \equiv \sum_{d}^{D} e_d + \sum_{d}^{D}\sum_{l\neq d}^{D} e_l \tau_{ld} \tag{8.3}$$

Let us assume that τ_{ld} is constant across activities d and l, so that $\tau_{ld}=R$ across all productive activities within the region. Since $\sum_{d}^{D} e_D$ is the *regional knowledge stock* (E), Equation (8.3) boils down to:

$$K \equiv E[1+(D-1)R] \tag{8.4}$$

According to Equation (8.4), the regional knowledge is a function of (1) the knowledge capital stock, (2) the number of technologies residing in the region, and (3) the *coherence* (R) among activities. If the bundle of activities residing within the region are characterized by a high degree of coherence ($R>0$), then the aggregate knowledge base increase with the *variety of technological competences* (D), weighted by their average relatedness. Conversely, if regional activities are featured by no coherence ($R=0$), then the regional knowledge base is equal to the knowledge capital stock. Therefore, the traditional approach to the computation of the knowledge base turns out to be a special case where $R=0$. Equation (8.4) can be approximated as follows:

$$K \cong EDR \tag{8.5}$$

Substituting Equation (8.5) in (8.1) we therefore get:

$$g_{i,t} = f(E_{i,t-1} D_{i,t-1} R_{i,t-1}) \quad (8.6)$$

In view of the arguments elaborated so far we are now able to spell out our working hypotheses. The generation of new knowledge is a core activity strategic for the competitive advantage of regional economies. Cross-regional differences in the development of technological knowledge provide thus a possible, although not exhaustive, explanation for differential growth rates (Fagerberg, 1987; Maleki, 2000). In line with a well established tradition of analysis we therefore expect *E* to be positively related to productivity growth.

The creation of technological knowledge is likely to exert a triggering effect on regional economic growth. Traditional analyses of the relationships between knowledge and growth has viewed the former as a bundled stock, i.e. a sort of black box the dynamics of which are rather obscure. Recent advances in the understanding of the cognitive mechanisms underlying the process of knowledge production allows for proposing that knowledge is the outcome of a combinatorial activity. Agents undertake their search across a bounded area of the knowledge landscape, so as to identify combinable pieces of knowledge. In other words, recombinant knowledge is the outcome of a local search process.

Knowledge structure may therefore be represented as a network, the nodes of which represent the combinable technologies, while links represent the actual combinations. Regional knowledge base turns out to be featured by a fairly heterogeneous structure, rather than a bundled stock. Due to the local character of search, the positive effects of knowledge on productivity which stem from the recombination of different technologies are more likely to occur in contexts where agents are able to combine together different and yet complementary technologies. Conversely, the presence of activities based upon weak complementarity of technological competences makes it difficult to implement effective knowledge production. In this case knowledge dynamics may hardly trigger regional growth. Therefore, in order to foster productivity growth, the internal structure of regional knowledge ought to be characterized by a high degree of complementarity across technologies. The specialization in technological activities undergoing organized search strategies is thus likely to trigger regional economic performances and as a consequence knowledge coherence (*R*) is expected to positively affect productivity growth.

Knowledge structure is not supposed to be stable over time. Changes may be brought about by trying new combinations among technologies or by introducing brand new technologies within regional competences. Variety may turn out to be a key resource to the creation of new knowledge, and therefore to economic development. It is indeed related to the technological differentiation within the knowledge base, in particular with respect to the diverse possible combinations of pieces of knowledge in the regional context. The localness degree of search implies that variety is likely to engender sensible results in terms of knowledge creation when such diverse technologies are somehow related one another. Within an established technological trajectory, the combination of technologies that are unrelated is less likely to enhance the process of knowledge creation, and hence it is not expected to contribute economic growth. The expectation about D therefore depends very much on the qualification of the variety of combined elements. Within contexts featured by organized search strategies within selected technological trajectories, related variety is likely to dominate over unrelated variety. The combination of a variety of related technologies is likely to exert a positive effect on knowledge production, and hence growth, while the combination of unrelated technologies is likely to exert a negative effect on knowledge production, and hence on regional growth.

8.3 Methodology

In order to investigate the effects of the properties of regional knowledge base on productivity growth, we first calculate an index of multifactor productivity (MFP).[5] To this purpose we follow a standard growth accounting approach (Solow, 1957; Jorgenson, 1995; OECD, 2001). Let us start by assuming that the regional economy can be represented by a general Cobb–Douglas production function with constant returns to scale:

$$Y_{it} = A_{it} C_{it}^{\alpha_{it}} L_{it}^{\beta_{it}} \tag{8.7}$$

where L_{it} is the total hours worked in the region i at the time t, C_{it} is the level of the capital stock in the region i at the time t, and A_{it} is the level of MFP in the region i at the time t.

Following Euler's theorem, output elasticities have been calculated (and not estimated) using accounting data, by assuming constant returns to scale and perfect competition in both product and factors markets. The output elasticity of labour has therefore been computed as the factor share in total income:

$$\beta_{i,t} = (w_{i,t} L_{i,t}) / Y_{i,t} \tag{8.8}$$

$$\alpha_{i,t} = 1 - \beta_{i,t} \tag{8.9}$$

Where w is the average wage rate in region i at time t. Thus we obtain elasticities that vary both over time and across regions.

Then the discrete approximation of annual growth rate of regional TFP is calculated as usual in the following way:

$$\ln\left(\frac{A_i(t)}{A_i(t-1)}\right) = \ln\left(\frac{Y_i(t)}{Y_i(t-1)}\right) - (1-\bar{\beta})\ln\left(\frac{C_i(t)}{C_i(t-1)}\right) - \bar{\beta}\ln\left(\frac{L_i(t)}{L_i(t-1)}\right) \tag{8.10}$$

The basic hypothesis of this chapter is that differences in regional growth rates are driven by the characteristics of regional knowledge bases. The increase in the knowledge stock and in the knowledge coherence is likely to positively affect productivity growth, while the effects of variety are likely to depend on the degree to which the diverse technological competences are related one another.

The test of such hypothesis needs to model the growth rate of MFP as a function of the characteristics of the knowledge base. Moreover, as is usual in this kind of empirical settings, we include in the structural equation also the lagged value of MFP, $\ln A_{i,t-1}$, in order to capture the possibility of mean reversion. Therefore the econometric specification of Equation (8.6) becomes:

$$\ln\left(\frac{A_i(t)}{A_i(t-1)}\right) = a + b\ln A_{i,t-1} + c_1 \ln E_{i,t-1} + c_2 \ln D_{i,t-1} + c_3 \ln R_{i,t-1} + \rho_i + \sum \psi t + e_{i,t} \tag{8.11}$$

where the error term is decomposed in ρ_i and $\Sigma\psi t$, which are respectively region and time effects, and the error component ε_{it}. Equation (8.11) can be estimated using traditional panel data techniques implementing the fixed-effect estimator. It relates the rates of productivity growth to the characteristics of knowledge base. However, one needs also to control for the impact on the one hand of agglomeration economies, on the other hand of changing regional industrial specialization, so as to rule out the possibility that such effects are somehow captured by the knowledge-related variables. In view of this, we can write Equation (8.11) as follows:

$$\ln\left(\frac{A_i(t)}{A_i(t-1)}\right) = a + b\ln A_{i,t-1} + c_1 \ln E_{i,t-1} + c_2 \ln D_{i,t-1} + c_3 \ln R_{i,t-1} + c_4 AGGL_{t-1} + c_5 LOQ_{t-1} + \rho_i + \sum \psi t + e_{i,t} \tag{8.12}$$

The knowledge-related variables variety and coherence (respectively D and R) are calculated according to the methodology described in Section 5.2. Productivity growth rates depend now not only on knowledge capital stock (E) and on the knowledge characteristics. Following Crescenzi *et al.* (2007), the effects agglomeration economies are captured by the variable $AGGL$, which is calculated as the (log) ratio between regional population and size (square kilometres). The changing specialization is instead proxied by LOQ, i.e. the location quotient for manufacturing added value.

8.3.1 Panel data and spatial dependence

The analysis of the effects of knowledge on productivity growth at the regional level calls for a special focus on the geographical attributes of such relations, i.e. on location aspects. Regional scientists have indeed showed that geographical proximity may affect correlation between economic variables.

While the traditional econometric approach has mostly neglected this problem, a new body of literature has recently developed, dealing with the identification of estimators able to account for both spatial dependence between the relationships between observations and spatial heterogeneity in the empirical model to be estimated. Former treatment of spatial econometric issues can be found in Anselin (1988), subsequently extended by Le Sage (1999).

The idea behind the concept of spatial dependence is straightforward. The properties of economic and social activities of an observed individual are likely to influence economic and social activities of neighbour individuals. Formally this relationship can be expressed as follows:

$$y_{i,t} = h(y_{j,t}),\ i = 1,\cdots,n,\ j \neq i \tag{8.13}$$

The dependence can therefore be among several observations. If this is the case, structural forms like Equation (8.12) are likely to produce a bias in the estimation results. There are different ways to cope with this issue. First, one may apply spatial filters to the sample data, so as to remove the spatial structure and then apply traditional estimation techniques. Second, the relationship can be reframed using a spatial error model (SEM), in which the error term is further decomposed so as to include a spatial autocorrelation coefficient. Third, one may apply the spatial autoregressive model (SAR), which consists of including the spatially lagged dependent variable in the structural equation.

We decided to compare the SAR and SEM models in order to have a direct assessment of the spatial dependence of productivity growth between close regions. However, most of the existing literature on spatial econometrics propose estimator appropriate for cross-sectional data. Given the panel data structure of our sample, we therefore follow Elhorst (2003) extending Equation (8.12) so as to obtain the SAR (Equation 8.14) and the SEM (Equation 8.15) specifications:

$$\begin{aligned}\ln\left(\frac{A_i(t)}{A_i(t-1)}\right) = \xi W \ln\left(\frac{A_i(t)}{A_i(t-1)}\right) + b \ln A_{i,t-1} + c_1 \ln E_{i,t-1} + c_2 \ln D_{i,t-1} + \\ c_3 \ln R_{i,t-1} + c_4 AGGL_{t-1} + c_5 LOQ_{t-1} + \rho_i + \sum \psi t + e_{i,t}\end{aligned} \tag{8.14}$$

$$\begin{aligned}\ln\left(\frac{A_i(t)}{A_i(t-1)}\right) = b \ln A_{i,t-1} + c_1 \ln E_{i,t-1} + c_2 \ln D_{i,t-1} + c_3 \ln R_{i,t-1} + \\ c_4 AGGL_{t-1} + c_5 LOQ_{t-1} + \rho_i + \sum \psi t + \varepsilon_{i,t} + \varphi_t\end{aligned} \tag{8.15}$$

$$\varphi_t = \delta W \varphi_t + \mu,\ E(\mu_t) = 0,\ E(\mu_t \mu_t') = \sigma^2 I_N$$

where ξ is referred to as spatially autoregressive coefficient and W is a weighting matrix. This latter can be defined either as a contiguity or as a normalized distance matrix. In the analysis that follows we chose the second alternative, by building a 19 × 19 symmetric matrix reporting the distance in kilometres between the city centres of the regional chief towns.

8.4 The data

In this chapter we investigate the relationship between productivity growth and regional knowledge in Italian regions.[6] The data we used have been drawn from two main sources. We employed data from the regional accounts provided by the Italian Institute of Statistics (ISTAT) to calculate the MFP index. We used real GDP (1995 constant prices) as a measure of regional output, regional labour income to compute the output elasticity of labour, regional employment level as a proxy for labour input, real gross fixed investments to derive capital stock (see Appendix 8.A).

To calculate the measures of regional knowledge base we employed an original dataset of patent applications submitted to the European Patent Office, as proxy of technological activities within manufacturing sectors. Each patent is assigned to a region, on the basis of the inventors' addresses.[7] Detailed information about the patents' contents has been drawn from the Thomson Derwent World Patent Index®. Each patent is classified in different technological field according to the Derwent classification. All technologies are covered by 20 subject areas designated as follows: classes A to M are in chemicals, P to Q refer to engineering, S to X refer to electrical and electronic. Each of the subject areas is in turn subdivided into three-digit classes.

We used the three-digit classification to calculate both knowledge coherence and information entropy. The decomposition of the entropy measure has been conducted by considering the subject areas as subsets, so as to obtain information entropy both "within" and "between" subject areas.

The initial patent dataset consists of 55,377 observations and 336 three-digit classes spread across 19 regions over the period 1979–2003. After the calculations we ended up with a vector of five knowledge variables, observed for each region over the period 1981–2002. Such vector has then been matched with the vector of regional productivity growth rates over the same period for the corresponding regions.

Table 8.1 and Table 8.2 provide the descriptive statistics for the set of variables used in the analysis and show general information about the various sampled regions. The sample is made of 19 Italian regions[8] and is characterized by a high degree of variance regarding both the knowledge variables and the growth rates of multifactor productivity.

In particular, from Table 8.2 seems to emerge an interesting pattern of geographical distribution for the knowledge variables. For example, while we expected negative values for knowledge coherence in North-Western regions, similar evidence for some North-Eastern regions is slightly puzzling. Negative values of knowledge coherence are indeed to be associated with periods of random screening in research activities, typical of exploration stages. Innovation

Table 8.1 Descriptive statistics

Variable		*Mean*	*Std. Dev.*	*Min*	*Max*	*Observations*
E	overall	1,232.625	2,380.950	1.000	15,795.300	N = 418
	between		1,979.379	29.605	8,106.422	n = 19
	within		1,391.848	–6,400.797	8,921.506	T = 22
R	overall	0.373	0.953	–0.545	6.407	N = 418
	between		0.671	–0.316	2.125	n = 19
	within		0.697	–2.243	5.041	T = 22
TV	overall	7.371	2.262	0	11.297	N = 418
	between		1.862	4.139	10.771	n = 19
	within		1.382	–0.086	9.884	T = 22
RTV	overall	2.525	1.293	0	5.178	N = 418
	between		1.129	0.839	4.649	n = 19
	within		0.703	–1.838	3.821	T = 22
UTV	overall	4.866	1.138	0	6.416	N = 418
	between		0.799	3.459	6.118	n = 19
	within		0.841	0.188	6.816	T = 22
dlogA/dt	overall	0.014	0.048	–0.203	0.292	N = 418
	between		0.009	0.000	0.037	n = 19
	within		0.047	–0.200	0.269	T = 22

Notes

E: knowledge capital; *R*: knowledge coherence; *TV*: information entropy; *RTV*: within-group information entropy; *UTV*: between-group information entropy; dlogA/dt: growth rate of multifactor productivity.

systems featured by the predominance of a mature paradigm are likely to undertake research efforts along a variety of paths, unless new profitable fields are sorted out, leaving room to the exploitation stage (and the consequent rise in knowledge coherence). The evidence for regions like Emilia Romagna and Tuscany suggests therefore that their industrial and technological development is more similar to that of North-Western regions than to that of North-East, maybe due to their faster growth patterns during the 1980s.

8.5 Empirical results

In order to assess the effects of knowledge coherence and variety on regional productivity growth, we carried out a fixed-effect panel data estimation of Equation (8.12), which is reported in Table 8.3. Different estimations are shown, in which we consider alternatively *TV*, *RTV* and *UTV*. The first column shows the results for the estimation including the measure of general technological variety. The results are quite in line with what expected according to our working hypotheses. First, cross-regional differences in the accumulation of knowledge capital stock matter in explaining productivity differentials, as is shown by the positive and significant coefficient on the variable *E*. Second, knowledge capital stock is not

Table 8.2 Regional decomposition of variables (1981–2002)

	E	*R*	*TV*	*RTV*	*UTV*	*dlogA/dt*
Piemonte	3,860.667	–0.316	10.097	4.340	5.756	0.007
Valle d'Aosta	29.605	2.125	4.703	1.232	3.459	0.003
Liguria	708.112	0.532	8.306	2.707	5.617	0.000
Lombardia	8,106.422	–0.232	10.772	4.651	6.117	0.016
Trentino Alto Adige	246.614	0.189	6.930	2.277	4.635	0.019
Veneto	2,088.573	–0.206	9.036	3.654	5.386	0.023
Friuli Venezia Giulia	834.670	–0.103	7.846	2.737	5.118	0.018
Emilia Romagna	2,993.007	–0.223	9.651	4.357	5.285	0.017
Toscana	1,219.773	–0.155	8.903	3.161	5.742	0.011
Umbria	175.860	0.253	6.676	1.948	4.766	0.003
Marche	355.378	0.036	6.856	2.31	4.555	0.019
Lazio	1,380.175	0.038	8.934	3.071	5.876	0.022
Abruzzo	414.795	0.921	6.161	2.306	3.828	0.025
Campania	260.018	0.357	6.965	2.026	4.997	0.011
Puglia	175.072	0.243	6.436	1.803	4.649	0.014
Basilicata	34.280	1.496	4.292	0.8581	3.326	0.042
Calabria	46.251	1.060	5.357	1.216	4.102	0.016
Sicilia	308.488	0.063	6.387	1.699	4.661	0.000
Sardegna	73.174	1.114	5.423	1.176	4.237	0.007

Notes

E: knowledge capital; *R*: knowledge coherence; *IE*: information entropy; *RTV*: within-group information entropy; *UTV*: between-group information entropy; dlogA/dt: growth rate of multifactor productivity.

sufficient to characterize the production of knowledge at the regional level. It is important to account also for qualitative changes in the knowledge base. In this direction, the internal degree of coherence of regional knowledge base exhibits a positive and significant coefficient. The more related the diverse technological activities carried out within the region, the higher the rates of productivity growth. Dynamic economies of scope are at stake as long as they are searched through the combination of close technologies. Finally, variety is a measure of how much the system is able to develop new technological opportunities, and eventually foster economic growth. As expected, the coefficient of *TV* is positive and significant. Regarding our control variable, it must be stressed that the proxy for agglomeration economies is not significant, while the location quotient for manufacturing activities is, as one could expect, negative and significant.

Column (2) reports the results for the estimation including *UTV*. Also in this case the coefficient for knowledge capital is positive and significant, like the one for knowledge coherence. Regarding variety, our estimations show that *UTV* is not likely to exert statistically significant effects on regional productivity growth. Also in this case the only significant control variable is the location quotient, which shows a negative sign.

The estimation in column (3) takes account of *RTV*. Differently from the other estimations, the coefficient for the lagged levels of productivity is now

Table 8.3 Panel data estimates of Equation (8.12)

	(1)	*(2)*	*(3)*	*(4)*
intercept	–0.212** (0.093)	0.203** (0.93)	–0.295*** (0.101)	–0.302*** (0.102)
$\log A_{t-1}$	0.0315 (0.022)	0.0223 (0.021)	0.041* (0.023)	0.0399* (0.022)
$\log(E)_{t-1}$	0.0212** (0.009)	0.028*** (0.009)	0.0185** (0.008)	0.0173* (0.010)
$\log(R)_{t-1}$	0.0878*** (0.035)	0.0792** (0.035)	0.0911*** (0.035)	0.0929*** (0.035)
$\log(TV)_{t-1}$	0.0153** (0.007)			
$\log(UTV)_{t-1}$		0.0007 (0.001)		0.0011 (0.002)
$\log(RTV)_{t-1}$			0.005** (0.002)	0.005** (0.002)
$\log(AGGL)_{t-1}$	–0.0007 (0.002)	–0.0018 (0.003)	–0.0012 (0.003)	–0.0012 (0.003)
$\log(LOQ)_{t-1}$	–0.1581*** (0.032)	–0.1506*** (0.032)	–0.1725*** (0.033)	–0.1743*** (0.033)
Regional dummies	Yes	Yes	Yes	Yes
Time dummies	Yes	Yes	Yes	Yes
Rsq	0.33	0.32	0.33	0.33
F	6.55***	6.33***	6.61***	6.37***
N	395	395	395	395

Notes

Dependent Variable: $\log(A_t/A_{t-1})$. *: $p < 0.1$; **: $p < 0.05$; ***: $p < 0.01$. Standard errors between parentheses.

(weakly) significant, and with positive sign. Regarding the effects of knowledge capital, the results are well in line with what we have seen so far. The coefficient is indeed positive and significant. The same applies to knowledge coherence. Not surprisingly, the coefficient for *RTV* is positive and statistically significant. This means that the positive effects observed in the case of *TV* is driven by *RTV*. Econometric results in column (4), where *UTV* and *RTV* are put together, are coherent with column (3). Knowledge coherence affects positively productivity growth, as well as knowledge capital. Again, only *RTV* appears to significantly affect productivity growth.

The results showed so far provide interesting evidence about the effects of regional knowledge base on productivity dynamics. However, recent advances in the analysis of spatial economic dynamics have pointed to the importance of proximity among economic agents. While the focus on the regional level does not allow for investigating this issue from a microeconomic viewpoint, nonetheless the presence of cross-regional external economies may cause a bias in the estimation using techniques that do not account for spatial dependence.

Table 8.4 reports the results from the econometric estimation of the SAR model (Equation 8.14). For the sake of homogeneity, different models have been estimated, including alternatively *TV*, *RTV* and *UTV*. As is immediately clear, the inclusion of the spatially lagged dependent variable changes our results only to a very limited extent. Let us start from column (1). First of all, the coefficient for the spatially lagged variable is positive and significant. The coefficients of both knowledge capital and knowledge coherence are significant and, as expected, positive. Interestingly enough, the coefficient for *TV* is no longer statistically significant. This might be explained by arguing that the positive coefficient of variety observed in the standard fixed-effects estimations, captures the effects of stimuli coming from outside the region. Regarding the control variables, it may be noted that the location quotient shows also in this case a negative and significant coefficient. Differently from the previous estimates, the coefficient for agglomeration is now negative and statistically significant. Such result also finds explanation in the peculiarity of industrial development paths followed by Italian regions.[9]

Columns (2) and (3) include respectively *UTV* and *RTV*. The results are fairly persistent, in that still knowledge capital and coherence are positive and significant, while neither of the two variety measures turn out to be significant. Once again, the spatially lagged dependent variable exhibits a positive and significant coefficient, while both the control variables negatively affect regional productivity growth. Finally, the estimation in column (4) includes related and unrelated variety together, providing results consistent with the previous estimations.

In order to check for the robustness of our results, we present in Table 8.5 the results for the estimation of the SEM model (Equation 8.15). The results are basically the same across the four models estimated, and are very coherent with the SAR estimations. The effects of variety are statistically significant in none of the models, while knowledge capital and knowledge coherence confirm to positively and significantly affect regional productivity growth. Both agglomeration

Table 8.4 Results for the estimation of Equation (8.14) (spatial auto-regressive model)

	(1)	*(2)*	*(3)*	*(4)*
$\log A_{t-1}$	–0.012 (–0.914)	–0.012 (–0.92)	–0.005 (–0.40)	–0.005 (–0.38)
$W[\log(A_t/A_{t-1})]$	0.188** (1.98)	0.188** (1.98)	0.190** (1.99)	0.190* (1.80)
$\log(E)_{t-1}$	0.0145** (1.99)	0.014*** (3.22)	0.006 (1.19)	0.006 (0.87)
$\log(R)_{t-1}$	0.081*** (2.36)	0.081** (2.28)	0.091*** (2.52)	0.091*** (2.51)
$\log(TV)_{t-1}$	–0.001 (–0.14)			
$\log(UTV)_{t-1}$		–0.0002 (–0.11)		0.003 (1.36)
$\log(RTV)_{t-1}$			0.003 (1.36)	0.0002 (0.147)
$\log(AGGL)_{t-1}$	–0.005*** (–4.15)	–0.004*** (–4.15)	–0.004*** (–4.22)	–0.004*** (–4.21)
$\log(LOQ)_{t-1}$	–0.131*** (–4.08)	–0.131*** (–4.09)	–0.143*** (–4.32)	–0.144*** (–4.31)
Regional dummies	Yes	Yes	Yes	Yes
Time dummies	Yes	Yes	Yes	Yes
Log-likelihood	653.18	653.17	663.4	654.07
N	395	395	395	395

Notes

Dependent variable: $\log(A_t/A_{t-1})$. *t* of Student between parentheses. *: $p < 0.1$; **: $p < 0.05$; ***: $p < 0.01$.

and the relative location quotient show negative and significant coefficients, supporting the relevance of the idiosyncratic features of regional development paths in Italy. Finally, the coefficient for spatial autocorrelation is positive and significant across all the models, corroborating the argument of cross-regional transmission of productivity gains.

Summing up, the check for spatial dependence has provided interesting results with respect to impact of knowledge characteristics on economic growth. In particular, the effects of knowledge coherence appeared to be pretty persistent and robust across the different specifications and the different estimators implemented. The variety of observed combinations instead appears to be somehow neutralized by the spatially lagged dependent variable. This result is not that obvious, and would deserve further investigation.

8.6 Discussion

The results obtained in this chapter open up a new path to the empirical analysis of the determinants of cross-regional growth differentials, with particular respect to the effects of knowledge creation. Moreover, the set of indicators we used in our analysis can be well used to explore the determinants of efficiency of knowledge production processes within a knowledge production function approach.

Besides the theoretical and methodological contribution, the analysis we carried out sheds a new light on the study of regional development in Italy, which has failed to apply the interpretative framework provided by the economics of innovation to investigate cross-regional differences in growth patterns. A bit of economic history is in order here to help clarifying this point.

In the 1950s most Italian regions were rural, and populated by a large share of small and medium-sized enterprises, as opposed to North-Western regions, specialized in manufacturing activities, carried out by large firms. Analysing the distribution of growth rates and structural change at the regional level in the period 1950–1970, the Ancona School identified and found clues to a successful diffusion process of manufacturing activities towards such rural regions in the North-East and eventually in Central Italy, along the Adriatic coast. For this reason they proposed grouping such regions into a larger macro-area which has been eventually called NEC (North-East-Centre).[10] At the same time, the growth of manufacturing industries was slowing down in the North-West, wherein the growth of business service industries was already *in nuce* (Pettenati, 1991; Fuà and Zacchia, 1983).

More recent evidence shows that the Italian economy has retained its delay in the industrialization process also during the last decades of the twentieth century. Previous analyses of the evolution of the regional specialization index in manufacturing sectors reveal that the geographical pattern has changed significantly over time. Indeed, the North-Eastern and Central regions are characterized by specialization indexes increasing over the period 1981–2001. It seems that at the turn of the century North-Eastern and Central regions are characterized by specialization indexes very close to (and in the some cases even higher than) the

Table 8.5 Results for the estimation of Equation (8.15) (spatial error model)

	(1)	*(2)*	*(3)*	*(4)*
$\log A_{t-1}$	–0.013 (–0.94)	–0.019* (–1.79)	–0.005 (–0.36)	–0.005 (–0.36)
$\log(E)_{t-1}$	0.016** (2.22)	0.009*** (3.29)	0.010* (1.68)	0.008 (1.17)
$\log(R)_{t-1}$	0.083*** (2.41)	0.033 (1.102)	0.092*** (2.58)	0.093*** (2.60)
$\log(TV)_{t-1}$	0.001 (0.160)			
$\log(UTV)_{t-1}$		–0.0002 (–0.54)		0.0006 (0.39)
$\log(RTV)_{t-1}$			0.003 (1.39)	0.003 (1.45)
$\log(AGGL)_{t-1}$	–0.006*** (–4.59)	–0.002*** (–2.55)	–0.006*** (–4.75)	–0.006*** (–4.69)
$\log(LOQ)_{t-1}$	–0.126*** (–4.02)	0.005 (0.52)	–0.138*** (–4.25)	–0.139*** (–4.27)
Spatial autocorrelation	0.50*** (6.82)	0.48*** (6.22)	0.51*** (7.12)	0.50*** (6.91)
Regional dummies	Yes	Yes	Yes	Yes
Time dummies	Yes	Yes	Yes	Yes
Log-likelihood	661.99	636.58	662.96	
N	395	395	395	395

Notes

Dependent Variable: $\log(A_t/A_{t-1})$. *t* of Student between parentheses. *: $p < 0.1$; **: $p < 0.05$; ***: $p < 0.01$.

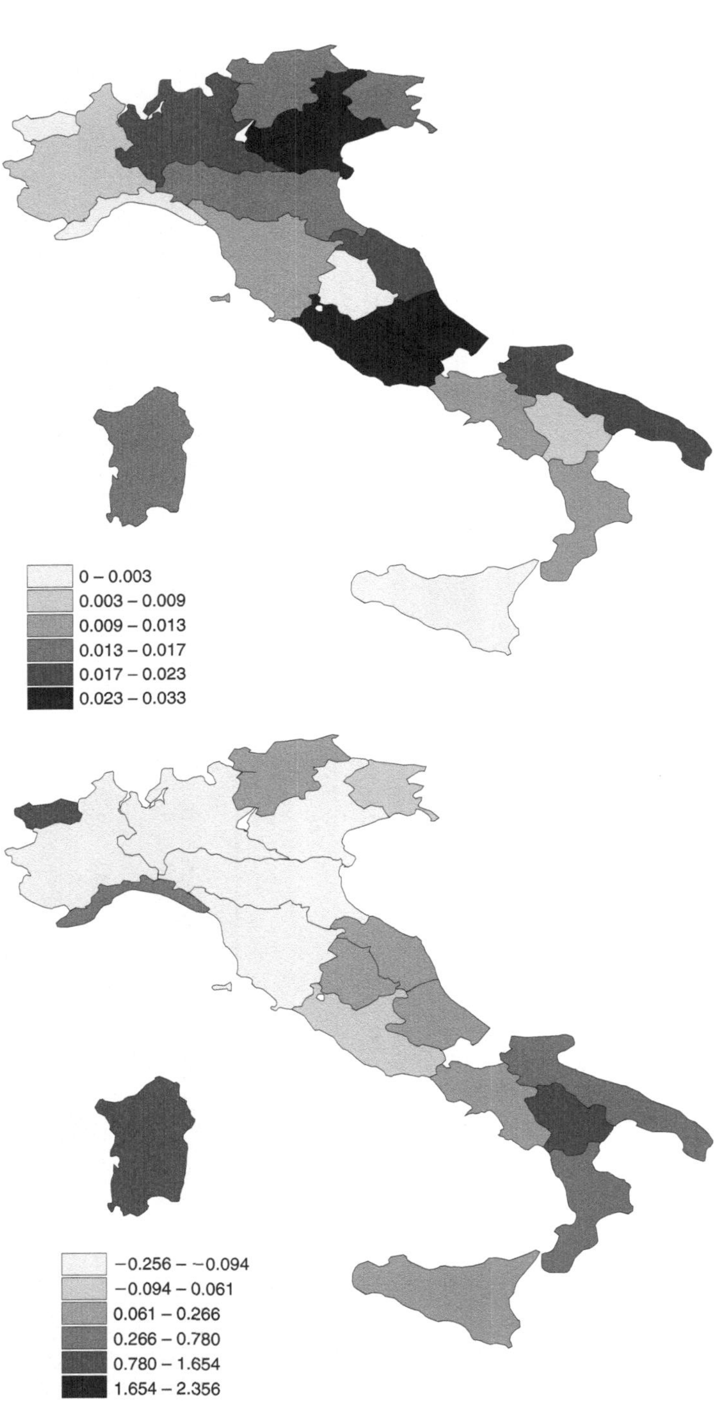

Figure 8.1 Cross-regional distribution of TFP and knowledge coherence.

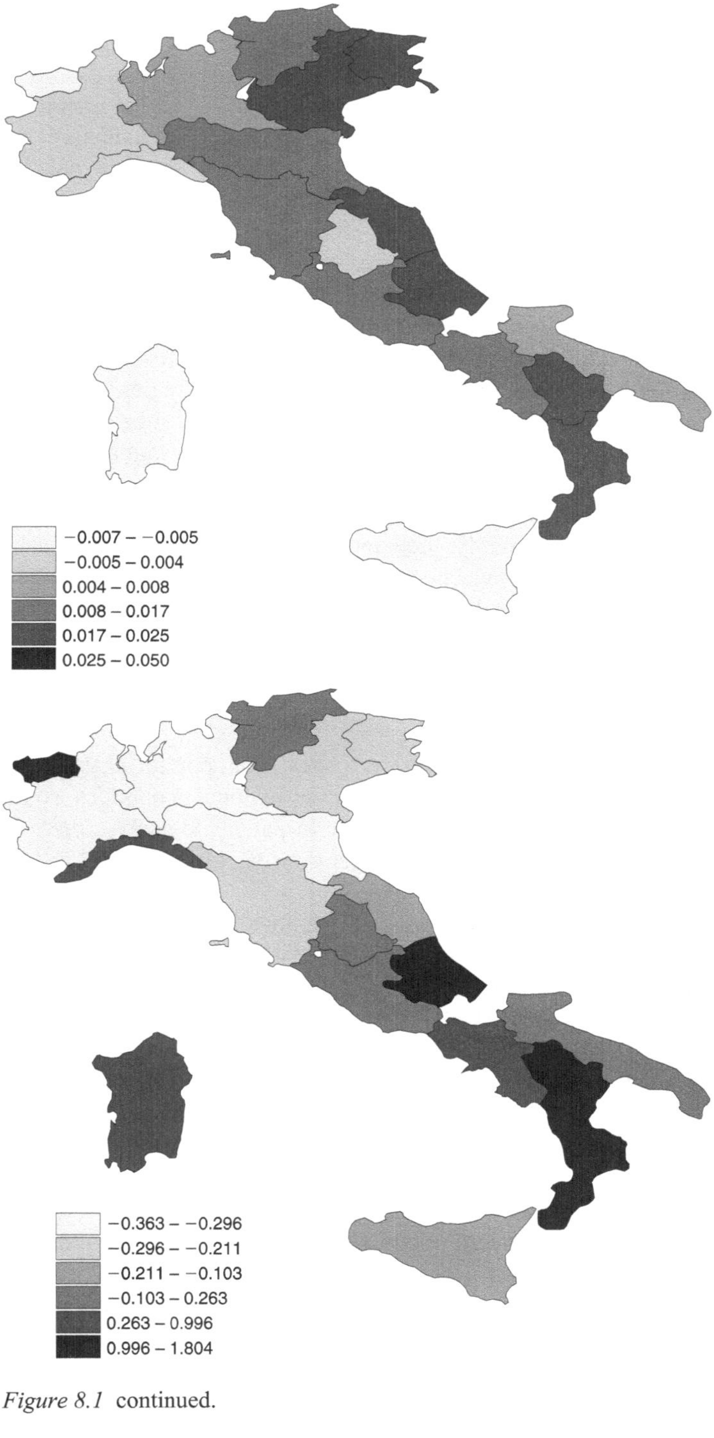

Figure 8.1 continued.

values featuring North-Western regions. Moreover, the trend appears to be soundly positive in the former, while the values in the latter are continuously decreasing since the early 1980s (Quatraro, 2009a, 2009b).

In this direction, the differential specialization of Italian regions in manufacturing sectors seems to produce diverse patterns of growth. The results of our analysis may contribute to better understanding this dynamics. With the help of Figure 8.1, we may argue that manufacturing sectors in early-industrialized countries have experienced the slackening of growth rates under the period of scrutiny, while early-industrialized regions, i.e. those in the North-East-Centre, have experienced increasing growth rates. Interestingly enough, these positive dynamics seem to have spread along the Adriatic coast to Southern regions. A look at the regional breakdown of knowledge coherence reveals how the index is pretty high in Central Italy and in the South. Of the North-East regions, the only one showing high values is the Trentino Alto Adige. This would suggest that the main prospects for growth for manufacturing industries are in lagging-behind regions. From a lifecycle perspective, latecomer regions seem to experience manufacturing-based growth dynamics that old industrialized regions experienced some decades ago. Accordingly, they appear to be in a phase of the technology lifecycle in which new knowledge is produced following rather organized search strategies. On the contrary, old industrialized regions face the major challenge to find out new avenues for boosting productivity growth rates. This involves exploration efforts in many possible directions, which look more like a sort of random screening wherein profitable new technologies still has to be found.

While the contribution to such a debate provides an important example of how this framework may be of interest to scholars in regional economics, some limits need to be discussed concerning (1) the use of patents to analyse innovation patterns on the one hand, and (2) the extension of Nesta's model to the regional domain on the other.

The use of patent applications as a proxy for innovation presents indeed a number of caveats which have already been discussed in Section 8.5. In addition, their use to analyse the Italian case might provide biased results, due to the size specialization of companies and to the existence of empirical studies emphasizing the scarce propensity of small firms to patent their innovations. It is indeed well known that about the 99 per cent of Italian firms are SMEs and this might lead to an underestimation of the phenomenon. However, the issue is far from a clear-cut solution. Empirical contributions in economics have indeed questioned the idea that small firms are more reluctant to innovate. For example, Brower and Kleinknecht (1999) emphasize that small firms develop larger portfolios of patent applications to counterbalance their lower market power. In addition, Lotti and Schivardi (2005) test the existence of a non-linear relationship between size and patent applications, suggesting that both small and large firms patent more than medium-sized ones.

For what concerns the second point, the regional extension of Nesta's model presents pros and cons deserving consideration. While the application of the framework at the firm level has the merit to stress and valorize the heterogeneous nature of firms' competences, an important limit can be identified in the focus on the firm as a single innovating agent, with no emphasis on cross-firm knowledge spillovers.

The shift to the regional domain is favoured by the consistency of the model with an interpretative framework blending the collective knowledge and the recombinant knowledge approaches. New knowledge stems out of a complex set of interactions among different institutions, of which firms represent only one out of different actors. Such interactions allows for the recombination of bits of knowledge that are fragmented and dispersed among the different agents (Hayek, 1945). The regional glance is thus more appropriate to grasp the local dimension of such dynamics (Antonelli *et al.*, 2011), so as to investigate the intertwining of the features of the topology of geographical and of knowledge spaces. The architecture of knowledge network, as proxied by the knowledge indicators we described in Section 8.4, proved to matter in shaping regional growth rates. In particular, the internal coherence of the regional knowledge base is positively related to productivity growth. This is because it is maintained that such index is likely to signal the transition towards a phase of organized search within regional industrial activities. The likelihood of generation of new useful knowledge is higher during this phase, and therefore one expects to also observe positive effects on production processes and hence productivity growth.

A problem might be raised by the framework we developed in this chapter, similar to the one we observed to affect Nesta's model. While the regional approach allows for accounting for the dynamics of inter-organizational knowledge flows within local contexts, it risks underestimating the important role of external knowledge as emphasized by Bathelt *et al.* (2004), who suggest that global pipelines add value to the local buzz by fuelling variety. However, this is not inconsistent with our approach and results. Indeed, while the implementation of spatial econometrics is motivated by the need to reduce the biases emerging when dealing with cross-regional analysis, it also allows us to appreciate and somehow to quantify the effects of productivity dynamics outside the region. By assessing the effects of neighbour regions' productivity we are able to account for the cross-regional effects of productivity enhancing factors, of which knowledge dynamics represent the main representatives in our model. The neutralizing effect of the spatial lagged dependent variable on technological variety provides support to this idea,[11] which deserves to be carefully analysed in future research.

8.7 Conclusions

Innovation and technological knowledge have long been considered as key elements triggering productivity growth. Empirical analyses of this relationship have emerged in the line of Zvi Griliches' extended production function, according to which knowledge is considered as an additional input in the traditional production function. In this framework knowledge is considered as a bundled stock, which has been operationalized by applying a sort of permanent inventory method to cumulate an innovation flow measure subject to a depreciation rate.

A step forward is represented by the studies introducing the knowledge production function. This strand of literature has mainly been developed to investigate innovation dynamics at the regional level. Drawing upon the regional innovation systems approach, it has basically provided a former empirical assessment of the degree to which knowledge is the result of the interaction of a number of different and yet complementary institutions involved in innovation activities, like firms, universities, R&D labs and so on (Cooke *et al.*, 1997; Antonelli, 2008).

While these studies inquired into the determinants of the effectiveness of knowledge production at the regional level, they said very little about the effects of knowledge on regional growth. Moreover, knowledge kept being represented as a bundled stock, although conceived as stemming from interactive dynamics.

In this chapter we have attempted to provide evidence of the effects of knowledge on regional growth by going beyond the traditional representation of knowledge found in literature. The recombinant knowledge approach and its cognitive underpinnings proved to be very fertile in this respect. Knowledge is understood as the result of the combination of bits of knowledge identified in the knowledge space by means of a local search process. This allows for representing the structure of knowledge as a web, the nodes of which are bits of knowledge, while the links stand for their actual combination. Such representation is susceptible of different operational translations. In this chapter we have followed the methodology elaborated by Nesta (2008), relying on information provided within patent documents.

We have grafted this methodology into an empirical framework analysing the effects of the characteristics of knowledge structure on regional productivity growth. Our analysis concerned a sample of 19 Italian regions over the period 1981–2002, focusing on manufacturing sectors. We have calculated annual multifactor productivity growth for each region, and then we have tested the explanatory role of knowledge variables such as the traditional knowledge capital, knowledge coherence and knowledge variety, both related and unrelated.

Summing up, the results of empirical analysis confirm that the regional knowledge base do affect productivity growth rates. In particular, not only the level of knowledge stock matters, but the characteristics of the knowledge base exert also a strong impact. The effects of variety are appreciable when spatial dependence is not accounted for. In particular, we decomposed total variety into related and unrelated variety. We have found that the positive effects of total variety are driven by related variety, while unrelated variety yields not significant effects. Regarding knowledge coherence, its effects are persistent and robust across all the alternative models and estimators implemented. The higher the internal degree of coherence of knowledge structure, the faster regional productivity is supposed to grow.

Such results have important policy implications, in terms of regional strategies for innovation and knowledge production. The internal coherence of the knowledge base proved indeed to positively affect productivity growth rates. Moreover, the specificity of the Italian case allows also for appreciating the importance of the relative maturity of the main industries, and the linkages between industrial and technology lifecycles. An effective regional innovation strategy should therefore be characterized by a careful assessment of local specificities. The identification of industries which the areas are specialized in is of paramount importance in order to devise the most appropriate incentive schemes. On the one hand, regions dominated by declining industries should be helped to find out new trajectories for development, trying and valorising the existing competences by directing search efforts towards complementary fields. On the other hand, in those regions featured by industries at the frontier, innovation policies might be much more directed towards the generation of incrementally new knowledge drawing upon exploitation strategies.

In conclusion, regional innovation policies should be characterized by intentional and careful coordination mechanisms, able to provide an integrated direction to research and innovation efforts undertaken by the variety of agents that made up the innovation system. The regional production system would then take advantage of a bundle of technological activities showing a high degree of coherence and therefore more likely to be properly absorbed and successfully exploited.

Appendix 8.A

In order to calculate the stock of fixed capital at the regional level, we follow the procedure set out by Maffezzoli (2006), which can be summed up as follows. The official procedure to compute the capital stock is the Permanent Inventory Method (PIM). We assume fixed expected service lives, simultaneous exit mortality patterns and linear depreciation. As a consequence, the real gross capital stock can be computed as:

$$\tilde{C}_t = \sum_0^{d-1} I_{t-i} \tag{A.1}$$

where d is the expected service life, and I_t the real investment flow at time t. The depreciation of capital stock is simply equal to $D_t = \tilde{C}_t / d$. The discrete approximation of such a relationship is:

$$D_t = (\tilde{C}_t + \tilde{C}_{t+1}) / (2d) \tag{A.2}$$

Finally the net capital stock obtains directly from $C_t = \Sigma_{i=0}^{d-1} I_{t-1}[1-(2i+1)/2d]$ or via the accumulation equation $C_t = C_{t-1} + I_t - D_t$.

The accounting data at the regional level provide series about gross fixed investments. To make calculations of regional capital stocks we drew the capital stock estimations and the depreciation data at the national level. Then we estimated the average expected service life of aggregated assets by rearranging Equation (A.2) as follows:

$$d = (\tilde{C}_t + \tilde{C}_{t+1}) / (2D_t) \tag{A.3}$$

The results suggest that the aggregate assets are expected to live on average about 34 years. Unfortunately the data about regional accounts are available only from 1980, so that we have not enough observation to compute the capital stock. We thus constructed a time series for the actual, time-varying and nationwide depreciation rate, defined as $\delta_t = D_t / K_{t-1}$, and then took 2001 as a benchmark starting point. We finally extended the series before and after 2001 using the following relationships respectively:

$$C_{i,t-1} = (C_{i,t} - I_{i,t}) / (1 - \delta_t) \tag{A.4}$$

$$C_{i,t} = (1 - \delta_t) C_{i,t-1} - I_t \tag{A.5}$$

This methodology has some drawbacks, like approximating a linear depreciation scheme with a geometric one, ruling out regional differences in depreciation rates and some necessary degree of measurement error. However, given the availability of the data, it provides a good approximation for the purposes of our work.

Appendix 8.B

Corrrelation matrix

	logA	*log(E)*	*log(R)*	*log(RTV)*	*log(UTV)*	*log(TV)*	*log(LOQ)*	*log(AGGL)*
logA	1							
log(E)	0.7156	1						
log(R)	–0.4004	–0.2319	1					
log(RTV)	0.2965	0.4975	–0.0499	1				
log(UTV)	0.5776	0.5557	–0.3998	0.1981	1			
log(TV)	0.3894	0.7237	–0.1246	0.649	0.2283	1		
log(LOQ)	0.5542	0.2602	–0.4208	0.0721	0.6105	–0.0219	1	
log(AGGL)	0.5493	0.6326	–0.1183	0.4171	0.2112	0.6627	–0.0398	1

Notes

1 See Chapter 6, note 1.
2 This chapter is based on Quatraro (2010).
3 Italian regions present pretty heterogeneous features from both the economic and the social viewpoint. The purpose of this chapter is to understand the extent to which differences in regional knowledge bases might be responsible for such economic variety. Of course, this implies that some other factors may interact in explaining the observed variety. The econometric model we will propose is meant to reduce the bias due to omitted variables and spurious relationships.
4 Thomas (1975) articulated the implications of Perroux' framework on regional economic growth using a product lifecycle perspective, wherein the saturation of product markets are the main thing responsible for the slowdown of growth rates and the quest for innovations aimed at opening new markets.

5 Some basic questions of course remain as to what interpretations to give to these kinds of index. While Solow (1957) associated TFP growth with technological advances, Abramovitz (1956) defined the residual as some sort of measure of ignorance. Nonetheless it remains a useful signalling device, in that it provides useful hints on where the attention of the analysts should focus (Maddison, 1987).

6 We acknowledge that the use of administrative regions to investigate the effects of knowledge creation represents only an approximation of the local dynamics underpinning such process. Indeed administrative borders are arbitrary, and therefore might not be representative of the spontaneous emergence of local interactions. It would be much better to investigate these dynamics by focusing on local systems of innovation. However, it is impossible to find out data at such a level of aggregation. Moreover, the identification of local systems involve the choice of indicators and threshold values according to which one can decide whether to unbundle or not local institutions. This choice is in turn arbitrary, and therefore it would not solve the problem, but it would only reproduce the issue at a different level. Thus we think that despite the unavoidable approximation, our analysis may provide useful information on the dynamics under scrutiny.

7 The assignment of patent to regions on the basis of inventors' addresses is the most widespread practice in the literature (see, for example, Maurseth and Verspagen, 2002; Henderson *et al.*, 2005; Breschi and Lissoni, 2009; Paci and Usai, 2009, to cite a few). A viable alternative may rest on the use of applicants' addresses, above all when the assessment of knowledge impact on growth is at stake (see Antonelli *et al.*, 2010). However, when the analysis is conducted at local level of aggregation, and the geography of collective processes of knowledge creation is emphasized, the choice of inventors' addresses remains the best one.

8 We left out the Molise region due to very low levels of multi-technology patents.

9 Population density is indeed likely to be higher in early-industrialized areas in the North-West, while late-industrialized regions in the so-called "third Italy" were characterized by lower population density due to diffusion of population across larger areas instead of concentration within metropolitan cities.

10 The grouping of Italian regions is as follows. North-West: Piedmont, Lombardy, Valle d'Aosta and Liguria. North-East: Veneto, Emilia-Romagna, Friuli Venezia-Giulia, Trentino Alto-Adige. Centre: Tuscany, Abruzzi, Marches, Lazio, Umbria and Molise. South: Campania, Apulia, Calabria, Basilicata, Sicilia and Sardegna.

11 The issue of knowledge flows incoming from far areas is more articulated, and difficult to address with the available data. Following Breschi and Lissoni (2001), we acknowledge that when knowledge is at stake, epistemic communities are likely to emerge wherein the effect of geographical distance is mitigated by cognitive proximity. To this purpose, finer-grained information on co-inventorship patterns would be necessary. However, this is beyond the scope of this chapter.

9 The co-evolution of knowledge and economic structure

Evidence from European regions

9.1 Introduction

The evidence provided so far has shown the usefulness of the methodologies we introduced in Chapter 5 to make operational the systemic approach to knowledge structure. In the previous chapters we have investigated the effects that changes in the architecture of knowledge may have on different aspects of the economy, say its performances or its evolutionary patterns. We have emphasized how detected changes in the architecture of knowledge structure can be interpreted in terms of changing search behaviours, which are likely to feature specific stages of technology lifecycles. We have introduced the distinction between random screening and organized search, which we have integrated with the well known distinction between exploration and exploitation, to the purpose of better grasping the dynamic interactions between the conduct of innovating agents and the emergent property, i.e. knowledge, arising from them.

In this chapter we go a step further by explicitly investigating the co-evolutionary patterns of knowledge and economic structure. The key message of this book lies indeed in the endogeneity of structural change. We have spelled out this principle as a general one, applicable to any "organization" in the socio-economic system, at whatever level of aggregation. The dynamic interactions among economic agents are likely to shape a wide array of structures, including the economic and the knowledge ones. Such structures are in turn likely to condition agents' behaviour. Obviously, as subsystems in a nested hierarchy, knowledge and economic structures are likely to shape each other, setting in motion a chain of dynamic feedbacks in which it is very difficult to single out the mechanisms of causation.

For this reason in this chapter we attempt to look at the relationships between changes in knowledge and economic structure without formulating any a priori assumption either on the form that they can take or on the direction of causality. To this purpose we will connect three different methodological approaches. The first is evidently a methodology described in Chapter 5 in order to represent the architecture of knowledge structure. In particular we will draw upon co-occurrences matrices to calculate coherence, cognitive distance and variety indicators. Second, we will provide a synthetic account of the change of economic structure by implementing a "shift-share analysis" in order to disentangle the contribution to (labour) productivity growth of within-sector productivity dynamics and of reallocation of labour force across the different sectors. Finally, the relationships between these two sets of indicators have been investigated by using a vector autoregression (VAR) model, which we estimated via "reduced form" applying the least absolute deviation (LAD) estimator due to the distributional properties of the variables.

The analysis is carried out on European NUTS II regions and provide an interesting insight into the dynamic feedbacks between economic and knowledge structure. While some relationships go in the expected direction, in some cases we are confronted with somewhat more articulated patterns that call for a finer-grained representation of search behaviours of innovating agents. The rest of the chapter is organized as follows. The next section elaborates a model and introduces "shift-share" analysis. Section 9.3 provides a description of the data used and outlines the econometric strategy. Section 9.4 presents the results of the estimations and discusses them in the light of the argument offered in Chapter 4. Finally, Section 9.5 provides some temporary conclusions.

9.2 A model for knowledge and economic structure: the shift-share analysis

A formal model linking the change of economic structure to that of knowledge structure can be easily derived by using a traditional Cobb–Douglas production function like the following;

$$Y_{i,t} = AC_{i,t}^{\alpha} L_{i,t}^{\beta} K_{i,t}^{\delta} \tag{9.1}$$

One can therefore see that production in a region i at time t can be represented by such kind of function, in which C stands for fixed capital, L stands for labour services and K stands for knowledge inputs. As usual, α, β and δ are the output elasticities of capital, labour and knowledge respectively. Following Nesta (2008), let us apply the decomposition of knowledge input as shown in Section 8.2, according to which:

$$K \cong EDR \tag{9.2}$$

where E is the traditional measure of regional knowledge capital stock, D measures technological variety while R represents the coherence of the regional knowledge base. Let us now substitute Equation (9.2) into (9.1) as follows:

$$Y_{i,t} = AC_{i,t}^{\alpha} L_{i,t}^{\beta} [E^{\omega_E} D^{\omega_D} R^{\omega_R}]_{i,t}^{\delta} \tag{9.3}$$

where ω_E, ω_D and ω_R are the weighted attributed to each of the three properties. By multiplying the exponent δ by such weights we obtain the following:

$$Y_{i,t} = AC_{i,t}^{\alpha} L_{i,t}^{\beta} E_{i,t}^{\theta_E} D_{i,t}^{\theta_D} R_{i,t}^{\theta_R} \tag{9.4}$$

Assume now that such production function is characterized by constant returns to scale in the traditional inputs capital and labour, such that:

$$\alpha + \beta = 1$$

By multiplying both sides of the equation by L^{-1} we obtain:

$$\left(\frac{Y}{L}\right)_{i,t} = A\left(\frac{C}{L}\right)_{i,t}^{\alpha} E_{i,t}^{\theta_E} D_{i,t}^{\theta_D} R_{i,t}^{\theta_R} \tag{9.5}$$

The left-hand side of this equation clearly is a labour productivity index. In order to investigate the relationship between the change in knowledge structure and change in economic performances we need to totally differentiate Equation (9.5) as follows:

$$\Delta\left(\frac{Y}{L}\right) = \Delta A \frac{\partial\left(\frac{Y}{L}\right)}{\partial A} + \frac{\Delta\left(\frac{C}{L}\right)\left(\partial\left(\frac{Y}{L}\right)\right)}{\partial\left(\frac{C}{L}\right)} + \Delta E \frac{\partial\left(\frac{Y}{L}\right)}{\partial E} + \Delta D \frac{\partial\left(\frac{Y}{L}\right)}{\partial D} + \Delta R \frac{\partial\left(\frac{Y}{L}\right)}{\partial R} \tag{9.6}$$

Now, after calculating all the derivatives on the right-hand side of Equation (9.6), and dividing both sides by (*Y*/*L*) we yield the following:

$$\frac{\Delta\left(\frac{Y}{L}\right)}{\left(\frac{Y}{L}\right)} = \frac{\Delta A}{A} + \alpha \frac{\Delta\left(\frac{C}{L}\right)}{\left(\frac{C}{L}\right)} + \theta_E \frac{\Delta E}{E} + \theta_D \frac{\Delta D}{D} + \theta_R \frac{\Delta R}{R} \tag{9.7}$$

Equation (9.7) relates the change in knowledge characteristics to the change in labour productivity. While this has proven to be a useful result, we still need to decompose "generic" labour productivity growth into the differential contribution provided by changing reallocation of employment across sectors, i.e. the most traditional utilization of the concept of structural change in economics.

To this purpose, the so-called shift-share analysis provides an interesting methodology that can be integrated in this framework with a few more passages. As noted by Houston (1967), the origins of shift-share analysis can be dated to the seminal work by Daniel Creamer (1942), although it did not attain great success at least until 1960, when Perloff *et al.* employed it as an analytical tool in their work *Regions, Resources and Economic Growth*. It has been mostly used to investigate and disentangle the compositional mix and the competitive position of regions in the face of observed changes in some relevant variables (Esteban, 1972, 2000). In this chapter we will follow the approach developed by

Fagerberg (2000), who decomposed labour productivity into three major components, i.e. the allocative, the productivity differential and the interaction between the two. We start by rearranging labour productivity as follows (region subscripts are omitted for the sake of clarity):

$$\frac{Y}{L}=\frac{\Sigma_j Y_j}{\Sigma_j L_j}=\sum_j\left[\frac{Q_j}{N_j}\frac{N_j}{\Sigma_j N_j}\right] \tag{9.8}$$

Labour productivity at the system level can therefore be decomposed in the contribution provided by labour productivity of each sector *j* as well as by share of sector *j* in total employment.

If we set:

$$P_j=\frac{Q_j}{N_j} \tag{9.9}$$

$$S_i=\frac{N_j}{\Sigma_j N_j} \tag{9.10}$$

then:

$$\frac{Y}{L}=\sum_j\left[P_j N_j\right] \tag{9.11}$$

The variation in labour productivity can therefore be expressed as follows:

$$\Delta\frac{Y}{L}=\sum_j[P_{j,t-1}\Delta S_j+\Delta P_j\Delta S_j+S_{j,t-1}\Delta P_j] \tag{9.12}$$

Equation (9.12) can therefore be expressed in growth rates by dividing it by (*Y*/*L*):

$$\frac{\Delta\left(\frac{Y}{L}\right)}{\left(\frac{Y}{L}\right)}=\sum_j\left[\frac{P_{j,t-1}\Delta S_j}{\left(\frac{Y}{L}\right)}+\frac{\Delta P_j\Delta S_j}{\left(\frac{Y}{L}\right)}+\frac{S_{j,t-1}\Delta P_j}{\left(\frac{Y}{L}\right)}\right] \tag{9.13}$$

The first term between parentheses is the contribution to productivity growth from *changes in the allocation of labour* between industries. It will be positive if the share of high-productivity industries in total employment increases at the expenses of industries with low productivity. The second term measures the *interaction between changes in productivity* in individual industries and *changes in the allocation of labour* across industries. It will be positive if fast-growing sectors in terms of productivity will also increase their share in total employment. The third term is the *contribution from productivity growth within industries*.

We can now substitute Equation (9.13) into Equation (9.7) to articulate in an explicit form the relationship between change in economic and knowledge structure:

$$\sum_j \left[\frac{P_{j,t-1}\Delta S_j}{\left(\frac{Y}{L}\right)} + \frac{\Delta P_j \Delta S_j}{\left(\frac{Y}{L}\right)} + \frac{S_{j,t-1}\Delta P_j}{\left(\frac{Y}{L}\right)} \right] = \frac{\Delta A}{A} + \alpha \frac{\Delta\left(\frac{C}{L}\right)}{\left(\frac{C}{L}\right)} + \theta_E \frac{\Delta E}{E} + \theta_D \frac{\Delta D}{D} + \theta_R \frac{\Delta R}{R} \tag{9.14}$$

Equation (9.14) provides a useful starting point to the elaboration of an empirical strategy for the assessment of the dynamic interactions between structural change in knowledge and the economy. However, as we moved from a Cobb–Douglas production function, one would think that the left-hand side of the equation is a function of the right-hand side, which would be clearly inconsistent with the main hypothesis of this book, according to which structures are endogenous and mutually interdependent. For this reason, we will use Equation (9.14) mostly as a hint rather than an indication of the functional form characterizing the relationship between knowledge and economic structure. The empirical analysis will be indeed carried out by adopting a somewhat less restricted approach which will be based on the application of vector autoregression (VAR) models, which we will describe in the next section.

9.3 Empirical approach

The main focus of this chapter is on the observation of the co-evolutionary dynamics between knowledge and economic structure. We have proposed in the previous section a synthetic representation of change in economic structure by introducing shift-share analysis. For the sake of clarity, let us assign a symbol to each of the identified components:

$$\mu = \sum_j \frac{P_{j,t-1}\Delta S_j}{\left(\frac{Y}{L}\right)} \tag{9.15}$$

$$\pi = \sum_j \frac{\Delta P_j \Delta S_j}{\left(\frac{Y}{L}\right)} \tag{9.16}$$

$$\alpha = \sum_j \frac{S_{j,t-1} \Delta P_j}{\left(\frac{Y}{L}\right)} \tag{9.17}$$

In view of the complex and endogeneous nature of the relationships between the properties of knowledge and those of economic structure, we apply a VAR model.

The regression of interest is the following:

$$w_{i,t} = c + \beta_z w_{i,t-z} + \varepsilon_{i,t} \tag{9.18}$$

Where w_{it} is an $m \times 1$ vector of random variables for region i at time t, β is an $\mathrm{m} \times [m \times z]$ matrix of slope coefficients that are to be estimated. In our particular case $m=9$ and corresponds to the vector [$\mu(i, t)$, $\pi(i, t)$, $\alpha(i, t)$, growth of knowledge capital (i, t), coherence growth (i, t), growth of cognitive distance (i, t), variety growth (i, t), related variety growth (i, t), unrelated variety growth (i, t)]. ε is an $m \times 1$ vector of disturbances. Knowledge capital is obtained by applying a permanent inventory method approach, the same way as the previous chapter. The properties of knowledge structure are instead calculated following the procedure described in Section 5.2.

In line with previous studies, the measure of growth rates is based on the difference of the logarithms of the respective variables. Let $X_i(t)$ represent the absolute value of the variable in region i at time t. Define the normalized (log) value of the variable as:

$$x_i(t) = \log(X_i(t)) - \frac{1}{N}\sum_{i=1}^{N} \log(X_i(t)) \tag{9.19}$$

here N is the number of regions. In what follows, growth rates are defined as the first difference of normalized (log) values according to:

$$g_i(t) = x_i(t) - x_i(t-1) \tag{9.20}$$

In such a way, common macroeconomic shocks are already controlled because the growth rate distribution was normalized to zero for each variable in each region in each year.

Following a growing body of literature (Coad, 2010; Buerger *et al.*, 2011; Colombelli *et al.*, 2011), Equation (9.20) is estimated via "reduced form" VARs, which do not impose any a priori causal structure on the relationships between the variables, and are therefore suitable for the purposes of this analysis. These reduced-form VARs effectively correspond to a series of *m* individual ordinary least squares (OLS).

However, previous studies have emphasized how the empirical distribution of the growth rates is closer to a Laplacian than to a Gaussian distribution (Bottazzi *et al.*, 2007; Bottazzi and Secchi, 2006; Castaldi and Dosi, 2009). Such evidence suggests that standard regression estimators, like OLS, assuming Gaussian residuals may perform poorly if applied to these empirical frameworks. To cope with this, a viable and increasingly used alternative consists of implementing the least absolute deviation (LAD) techniques, which are based on the minimization of the absolute deviation from the median rather than the squares of the deviation from the mean.

It must be noted that we do not include any individual dummies in the analysis. Even though unobserved heterogeneity can have important effects on the estimation results, the inclusion of individual dummies along with lagged variables may engender some biases for fixed-effect estimation of dynamic panel-data models, a problem known as Nickell-bias. Some alternative approaches relate to the use of instrumental variable (IV) or GMM estimators (Blundell and Bond, 1998). The main problem with this lies in the difficulty to find out good instruments, which is particularly hard when dealing with growth rates. When instruments are weak, IV estimation of panel VAR thus leads to imprecise estimates. Binder *et al.* (2005) propose instead a panel VAR model including firm-specific effects, which is, however, based on the assumption of normally distributed errors, which is not the case for what concerns the growth rates of the variables used in our regressions.

Since we are dealing with growth rates, instead of levels, we can maintain that any region-specific component has been largely removed. Moreover, we follow the wide body of literature on the analysis of firms' growth rates stating that the non-Gaussian nature of growth-rate residuals are a far more important econometric problem deserving careful attention even in regional level analyses (Buerger *et al.*, 2011).

9.3.1 The data

In order to implement the analysis outlined in the previous section we gather together two datasets. The shift-share analysis has been conducted by using the branch accounts of NUTS II European regions[1] provided by the Eurostat within the European System of Integrated Economic Accounts. As is well known, these data are available only since 1995, the year in which the Euorostat has implemented a standardized procedure to collect data from European countries, so as to build a coherent and homogeneous dataset. As a result we were able to calculate the μ, the π, and the α components for a subset of European regions in the period 1995–2007. The properties of knowledge structure, i.e. coherence, cognitive distance and variety (based on the information entropy index) have instead been calculated by using patent information contained in the OECD REGPAT database which covers patent data that have been linked to regions utilizing the addresses of the applicants and inventors. The analysis has been conducted by adopting the inventor-based regionalization,[2] and by using four-digit technology codes.

We obviously merged the two sets of indicators on the basis of the NUTS II regional code and the year. We end up with an unbalanced panel of 227 firms observed on average on eight years. The descriptive statistics for the whole sample are reported in Table 9.1, while Figure 9.1 shows instead the distributional properties of the variables under scrutiny, providing empirical support to their non-Gaussian distribution. In particular all the variables appears to follow a Laplace-like distribution, which makes our empirical strategy outlined in the previous section the best approach to the analysis.

Table 9.1 Descriptive statistics of the nine variables before normalization

Variable		*Mean*	*Std. Dev.*	*Min*	*Max*	*Obs.*
μ	overall	0.003	0.006	–0.030	0.060	N = 1,876
	between		0.005	–0.030	0.033	n = 227
	within		0.005	–0.028	0.042	
π	overall	–0.001	0.002	–0.032	0.002	N = 1,873
	between		0.002	–0.022	0.001	n = 227
	within		0.001	–0.023	0.008	
α	overall	0.024	0.068	–0.349	0.630	N = 1,873
	between		0.044	–0.217	0.173	n = 227
	within		0.061	–0.361	0.507	
Knowledge capital	overall	0.048	0.109	–0.163	1.253	N = 1,711
	between		0.102	–0.101	0.644	n = 202
	within		0.090	–0.659	0.756	
Knowledge coherence	overall	–0.004	0.032	–0.644	0.578	N = 1,744
	between		0.046	–0.159	0.578	n = 205
	within		0.026	–0.489	0.239	
Cognitive distance	overall	0.001	0.115	–1.099	1.099	N = 1,744
	between		0.056	–0.409	0.549	n = 205
	within		0.111	–1.098	1.107	
Knowledge variety	overall	0.003	0.064	–0.333	0.568	N = 1,744
	between		0.069	–0.145	0.494	n = 205
	within		0.052	–0.294	0.393	
Related knowledge variety	overall	0.004	0.113	–0.601	1.615	N = 1,744
	between		0.115	–0.601	0.726	n = 205
	within		0.095	–0.664	0.892	
Unrelated knowledge variety	overall	0.005	0.103	–1.163	1.125	N = 1,742
	between		0.094	–0.519	0.518	n = 205
	within		0.090	–0.728	0.998	

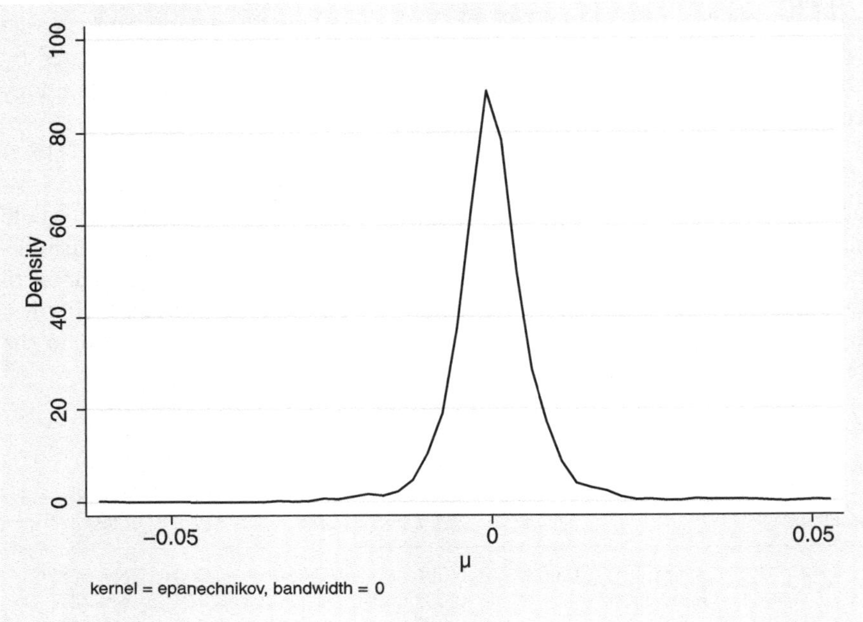

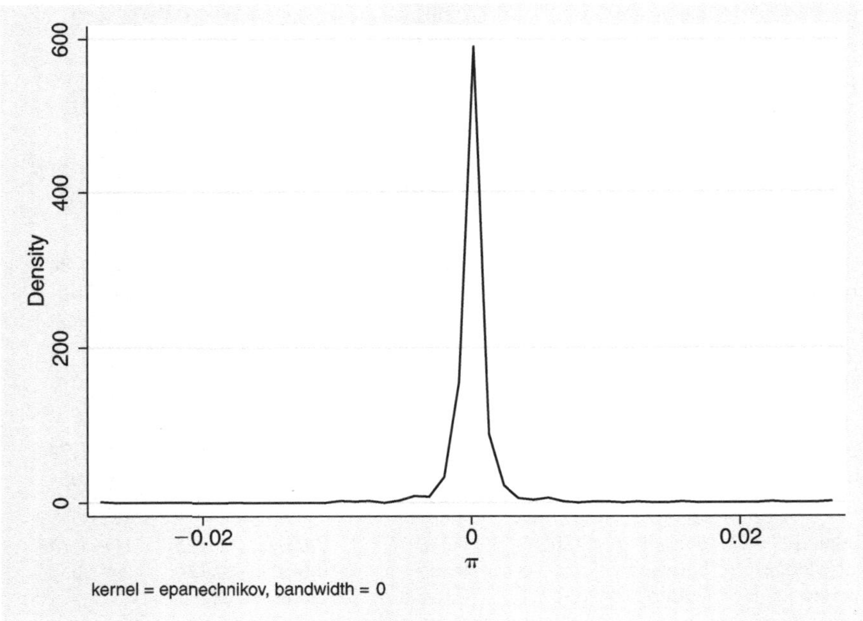

Figure 9.1 Distribution of the nine relevant variables describing knowledge and economic structure.

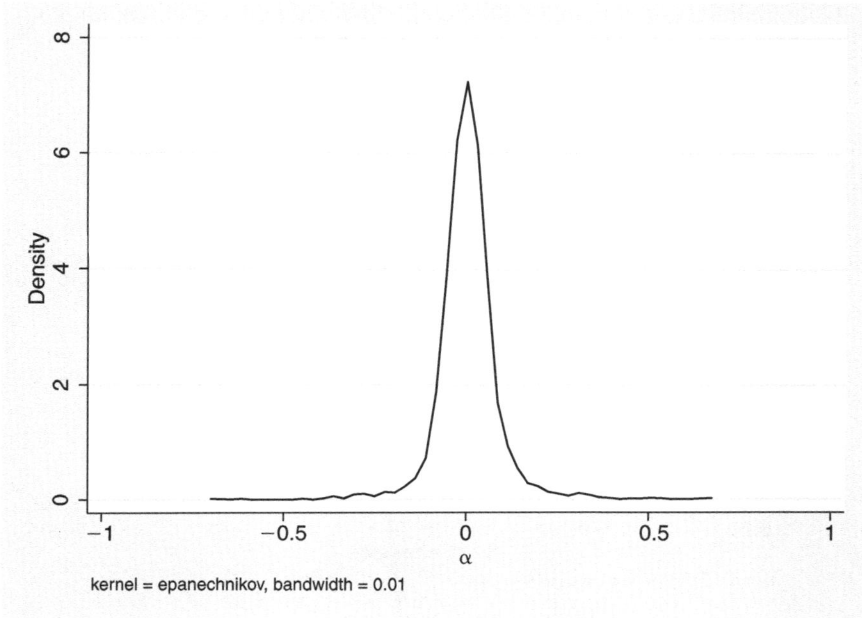

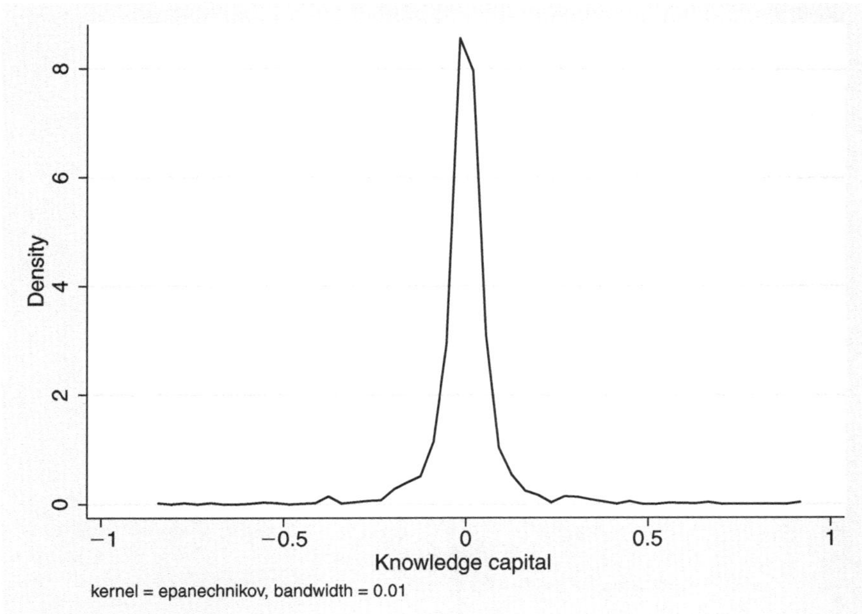

Figure 9.1 continued.

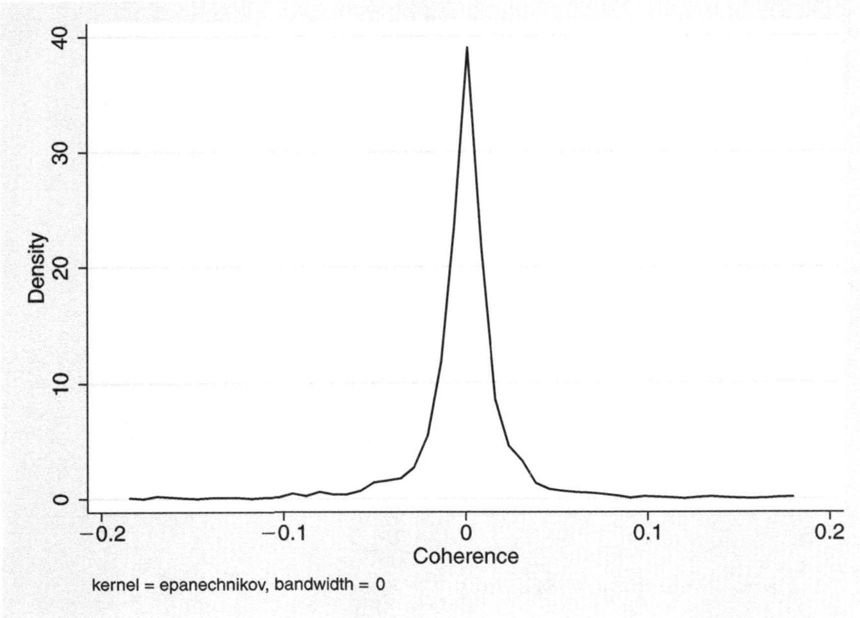

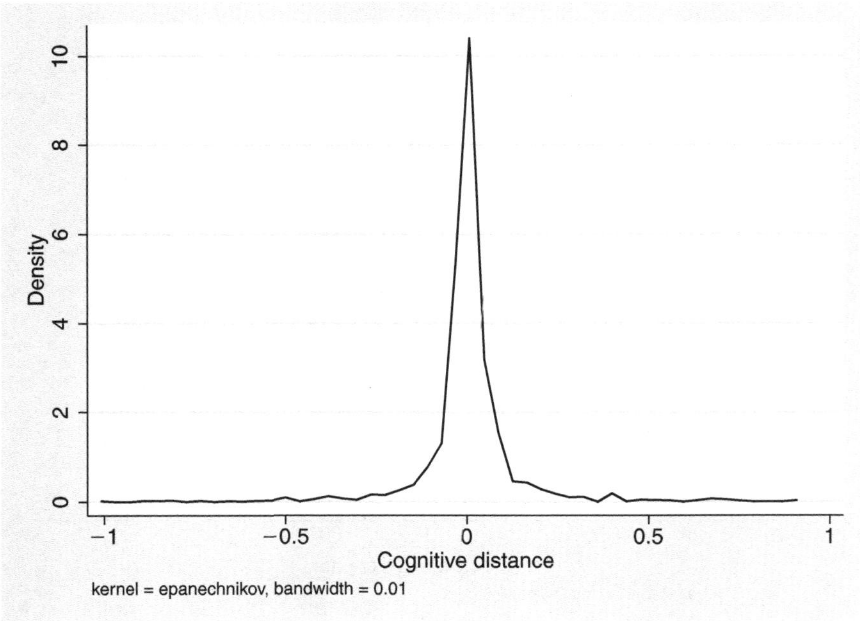

Figure 9.1 continued.

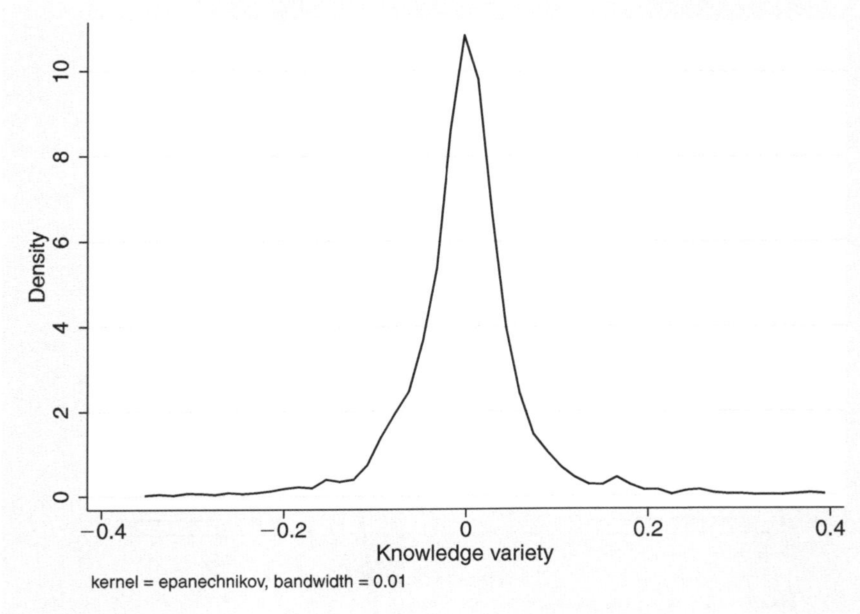

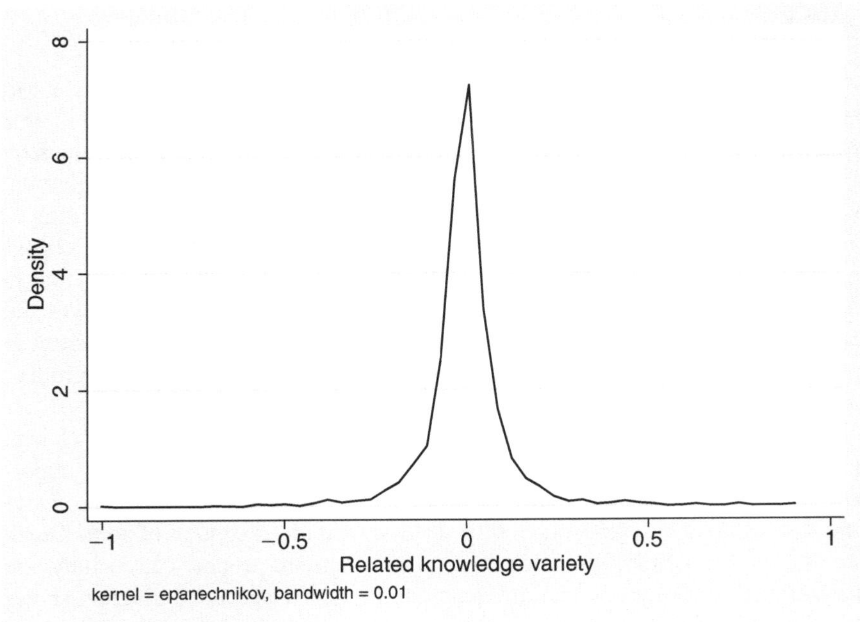

Figure 9.1 continued.

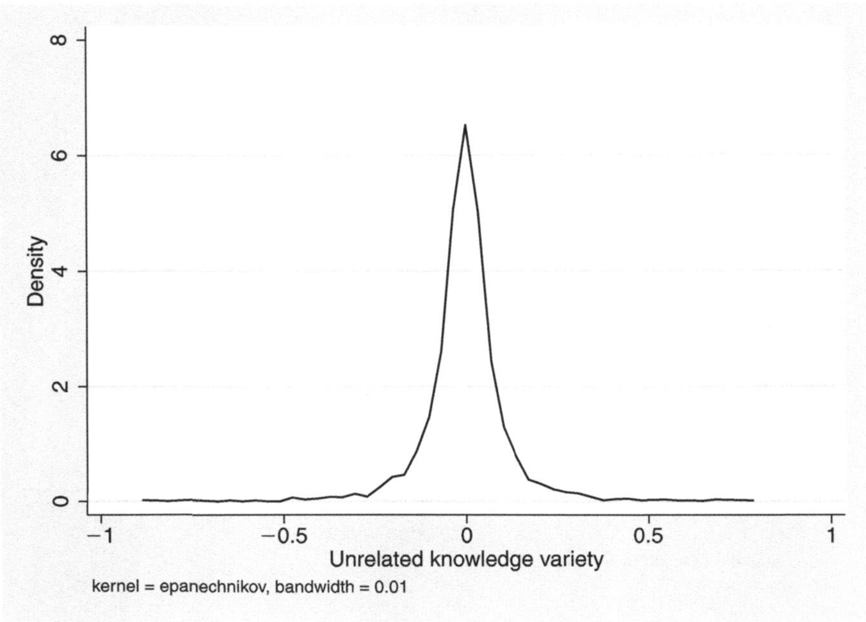

Figure 9.1 continued.

The elaboration of a regional breakdown of descriptive statistics turns out to be very much complicated when dealing with a sample of 227 firms. For this reason we decided to show the cross-regional distribution of average values by implementing a map for each of the variables under consideration. In Figure 9.2 we report the cross-regional distribution of the three components contributing to labour productivity growth. Let us recall that μ is the contribution of the changing mix of regional industries, and is positive if regions tend to specialize in high-productivity activities, π is the interaction between productivity growth and the change in the industry mix, and is positive to the extent that regions specialize in fast growing sectors, while α is the contribution of within-sector productivity growth weighted by the sector share on total employment.

It is interesting to note that for most of sampled regions the effect of change in the industry mix is positive, suggesting that structural change plays an important role in the process of economic growth. Most European regions tend therefore to specialize in high-productivity sectors, with the exception of some Greek regions and the British Midlands. The process is more pronounced in Italy and in Central-Eastern Europe than in Spain and France. The second diagram shows that the interaction term is positive again in most Italian regions, Spain, France and Germany, while the evidence is more mixed in the other regions. Italy, France, Spain and Germany in the observed period are subject to changes favouring the increasing share of fast-growing sectors. Finally, the within-sector

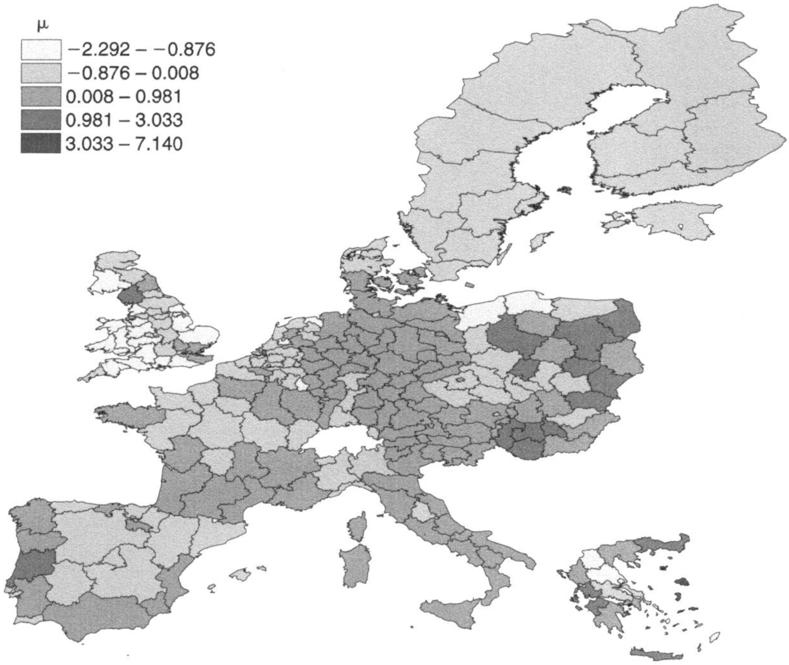

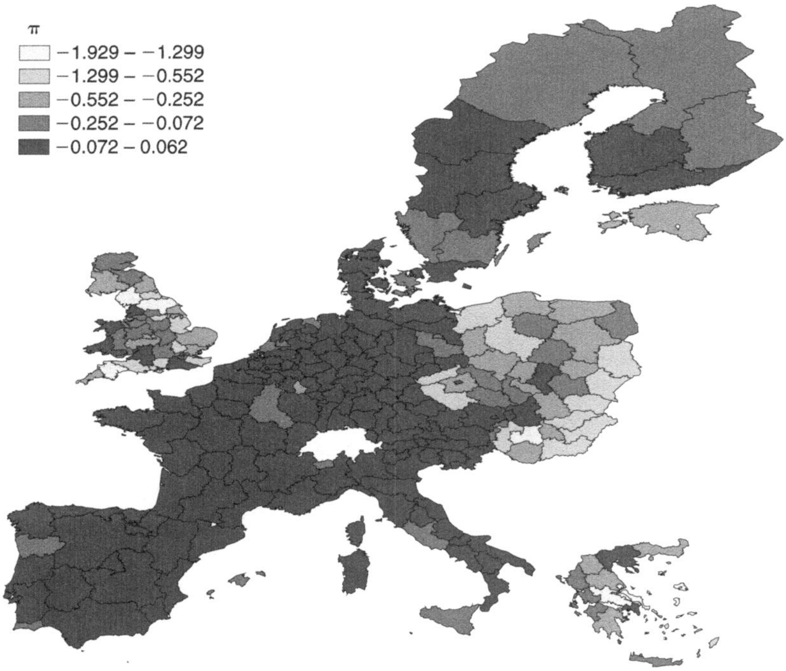

Figure 9.2 Distribution of the three components of shift-share decomposition.

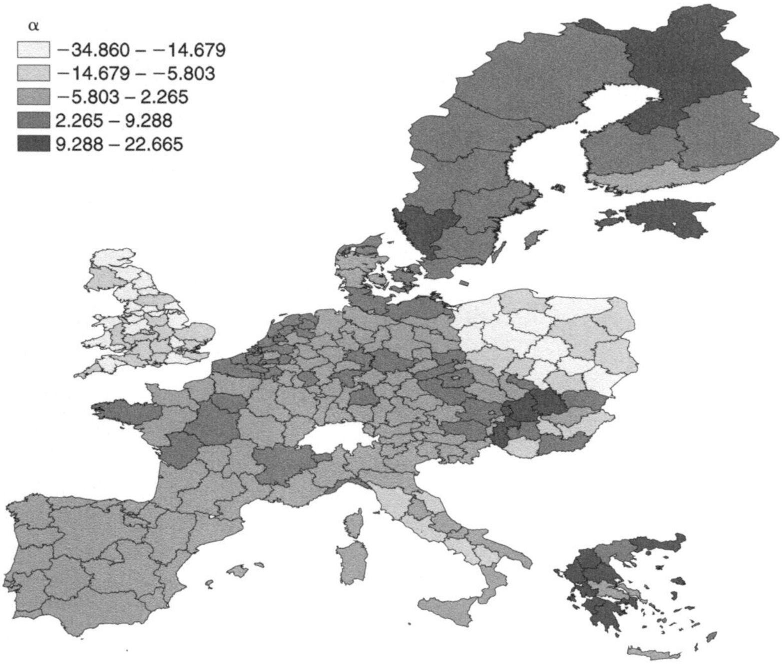

Figure 9.2 continued.

productivity growth seems to matter the most for northern regions, like Finland, Sweden and Denmark, and at a somewhat lesser extent for some eastern and Greek regions.

In Figure 9.3 we report instead the cross-regional distribution of knowledge capital, coherence and cognitive distance (log values). The top diagram reports the figures concerning knowledge capital. We can see how knowledge capital is higher in Central European regions and in northern regions, while it is lower in the periphery of the continent. A look at the coherence index reveals that on average search behaviours are more like organized search than random screening, while cognitive distance is on average very low in most of the European regions, suggesting that exploration is conducted across the safe boundaries of established knowledge competences. Only for a few scattered regions in France, Spain and Finland we observe low values of both coherence and of cognitive distance, suggesting a search strategy characterized by exploration behaviours conducted within well defined boundaries of the knowledge space.

In Figure 9.4 we show the cross-distribution of the variety index, articulated in unrelated and related knowledge variety. The top diagram indicates that on average European regions are characterized by a high degree of variety, with the only exceptions being peripheral regions in Portugal and in Greece. When we look at the distinction between related and unrelated variety we notice that the

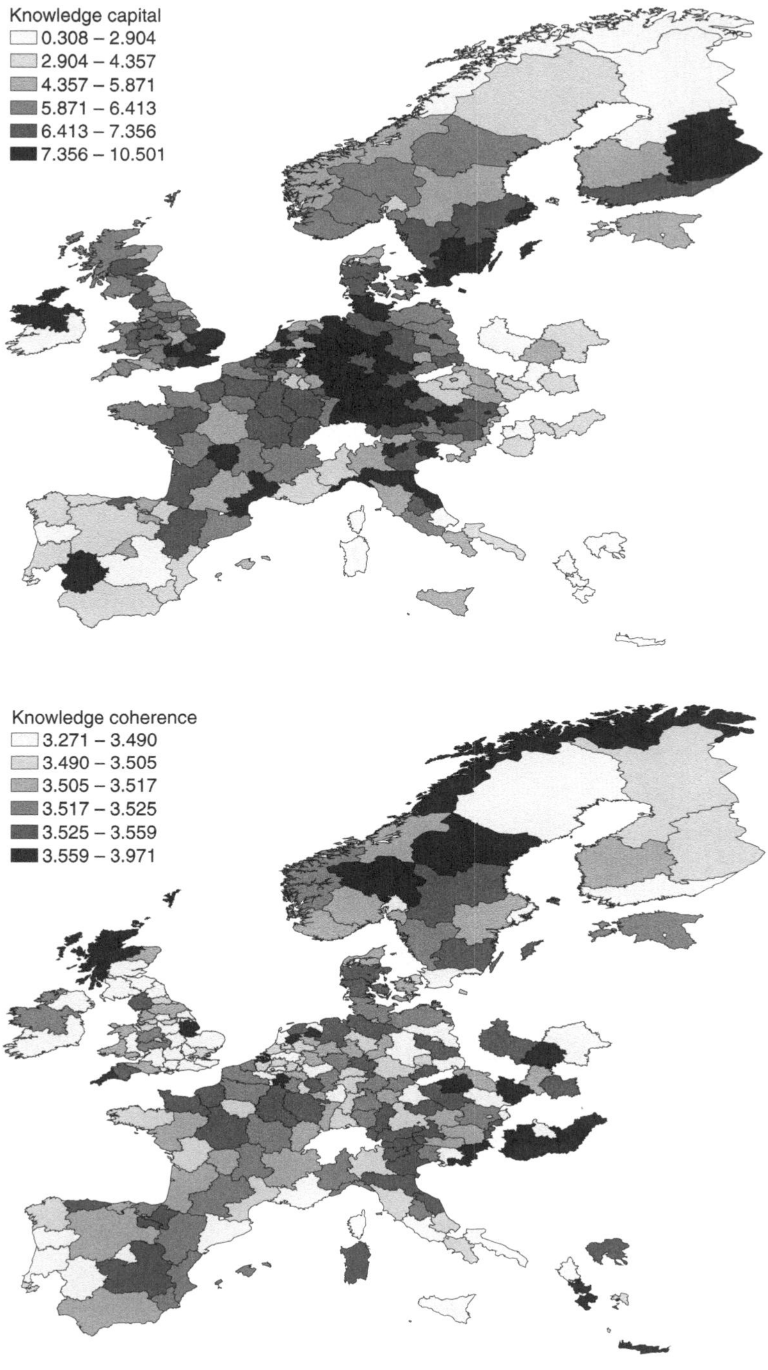

Figure 9.3 Distribution of the properties of knowledge structure (I).

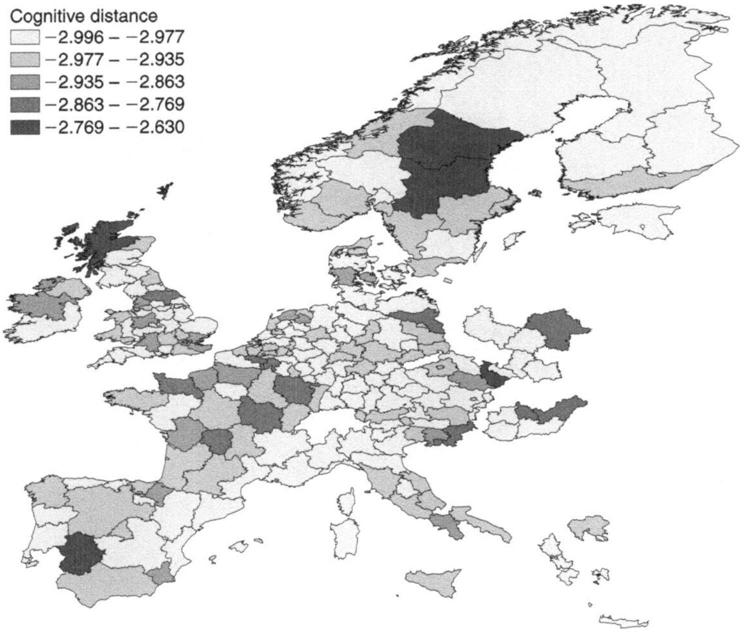

Figure 9.3 continued.

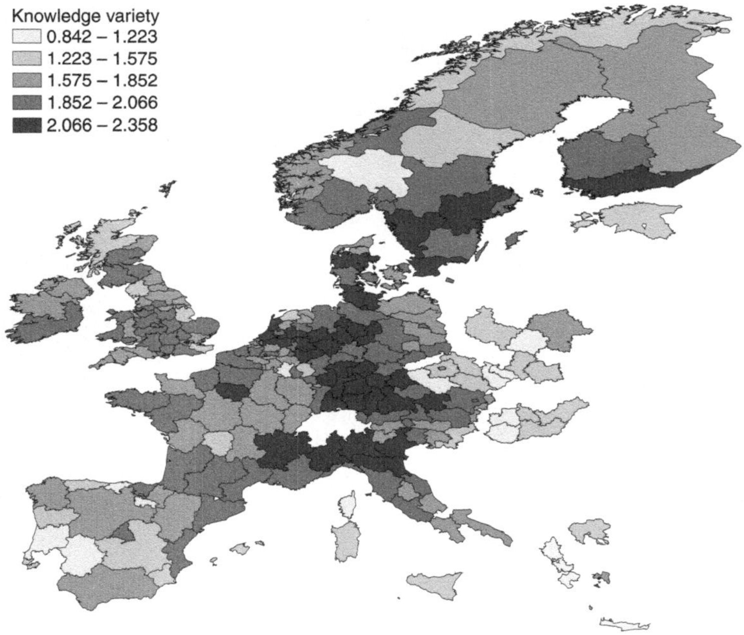

Figure 9.4 Distribution of the properties of knowledge structure (II).

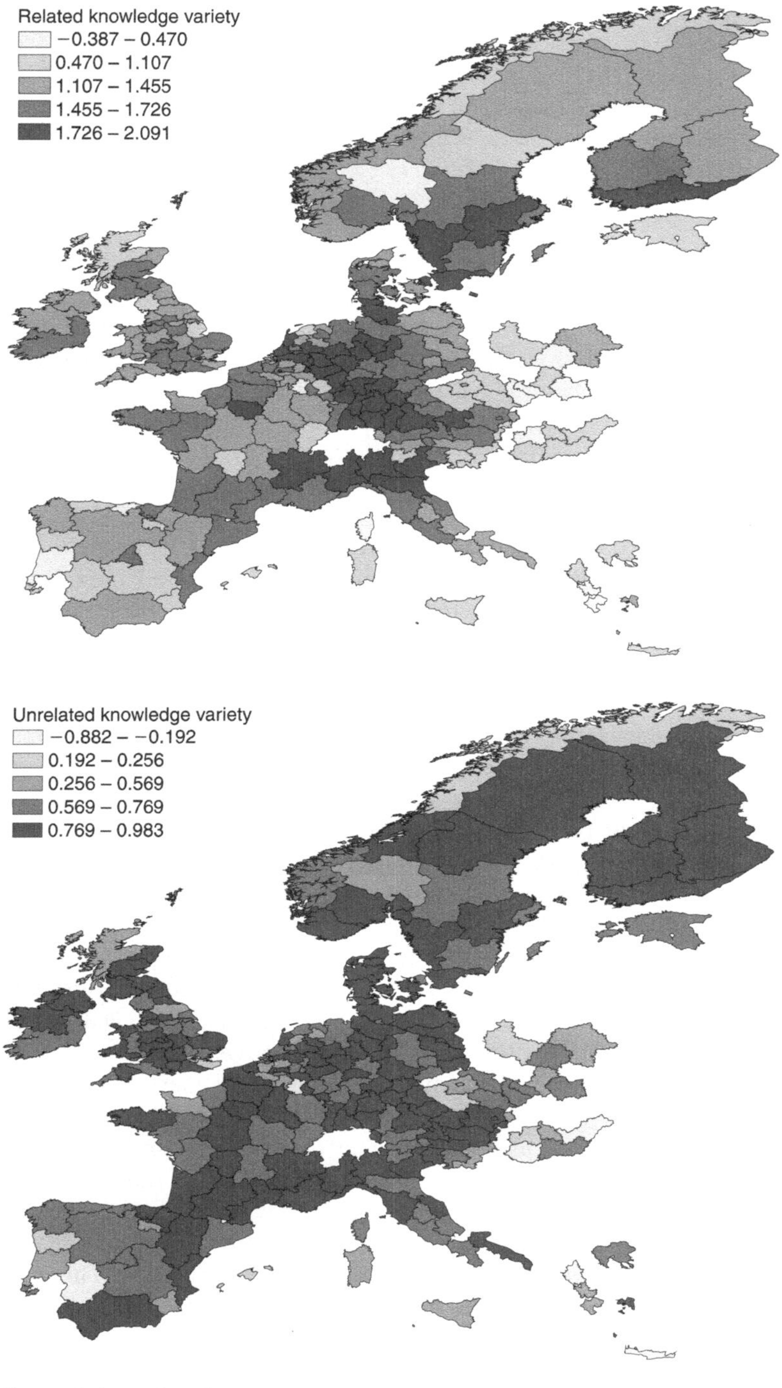

Figure 9.4 continued.

distribution looks very similar to that of total variety. By observing also the ranges assigned to each classes, we can also emphasize that on average related knowledge variety is higher than unrelated variety.

The maps reported in Figure 9.3 and in Figure 9.4 are based on absolute (log) values of the properties of the knowledge structure. In the following section we will implement the estimation of Equation (9.18), which is based instead on the normalized growth rates of such variables.

9.4 Econometric results

The results of the "reduced-form" VAR are reported in Table 9.2, which should be read as follows. Each column corresponds to each of the dependent variables in the model. Thus in column (1) the dependent variable is the normalized growth of μ, in column (2) that of π, and so on and so forth. The rows indicate instead the explanatory variables, which are grouped by lag (three lags are included). At the end of the table we report also the number of observations and the R-squared for each regression.

With respect to the observed autocorrelation, it is impressive that none of the variables under scrutiny shows any degree of persistence. On the contrary, coefficients are negative and significant across all the three lags considered, suggesting erratic growth dynamics for all the variables. Such results on knowledge-related variables are consistent with the findings of Buerger *et al.* (2011), who ascribe this kind of evidence to the intrinsic uncertainty and volatility characterizing innovation. Evidently, even though we attempted to counterbalance such volatility by letting each paper last five years, this has not been enough.

We now move to analyse in more detail the lead–lag relationship between the change in knowledge and in economic structure. As far as the first lag is concerned, knowledge coherence and knowledge capital show a positive and significant coefficient on α, which is consistent with the largest part of the literature linking knowledge and productivity growth. The α component stands indeed for the contribution stemming from within-sector productivity growth, which is positively affected by the growth of knowledge coherence and that of knowledge capital. Cognitive distance is instead negatively linked to π. The search across dispersed area of the knowledge landscape is therefore likely to jeopardize the increase in the share of fast-growing sectors. The knowledge variety indexes do not seem to affect significantly the economic structure. Unrelated variety appears instead to be positively affected by π, suggesting the increasing share of fast-growing sectors is likely to favour the introduction of further variety in the innovation system. It is also interesting to note that α affects negatively the growth of coherence and cognitive distance. This evidence is in line with previous work (Colombelli *et al.*, 2011), according to which higher performances are likely to created the economic conditions to stimulate exploration activities, although in domains that are not too far from the established technological competences.

When we move to the second lag, we see that knowledge coherence affects positively and significantly π, i.e. faster growth of coherence is associated with the faster increase of faster-growing sectors. Cognitive distance is again negatively related to π, while the related and unrelated variety indexes are instead both related positively to π. This evidence is quite puzzling, as by definition when related variety rises, unrelated variety decreases. However, the twin positive coefficients can be interpreted in the light of the mixed nature of the π component, whereby related variety positively affect productivity growth, while unrelated variety positively affects the change in the industry mix. Regarding the effects of the economic structure on knowledge structure, the μ component does not yield any significant effect on the knowledge characteristics. The π component instead affects positively coherence and negatively cognitive distance: the increasing share of faster-growing sectors stimulates the establishment of exploitation activities dominated by organized search strategies within the comfortable fences of established competences. Once again, α negatively affects knowledge coherence and cognitive distance, like in the one-lag coefficient.

Finally, the third lag presents an interesting negative and significant coefficient on the effect of knowledge coherence on μ, which suggests that the decrease of knowledge coherence, which signals the undertaking of exploration activities, is likely to engender a reallocative effect of labour force across sectors, i.e. to foster the change in economic structure. The effect on π is again positive, signalling the prevalence of the positive effects on productivity dynamics. The coefficient of cognitive distance on π is instead negative and significant, which, coupled with the positive one of coherence, suggests that exploitation strategies based on organized search are likely to engender the movement towards faster-growing activities. Regarding the effects of economics on knowledge structure, both π and μ yield negative and significant effects on knowledge variety, and in particular on related variety. Thus it would seem that increasing variety fosters the changing allocation of labour across sectors, but that this in turn is likely to be followed by a reduction in variety. The convergence towards faster-growing sectors is also followed by a sharp decrease of cognitive distance. The within-sector productivity dynamics do not seem to hold significant effects on knowledge structure.

9.5 Conclusions

In this chapter we have conducted an exploratory analysis of the co-evolutionary patterns of knowledge and economic structure. Drawing upon a theoretical framework which stresses the dynamic nature of the interactions between these two components as well as the endogenous character of their change process, we decided to implement an empirical framework based on the indicators proposed in Section 5.2 in order to characterize the architecture of knowledge structure. We coupled such methodological approach with the shift-share technique, which allows us to grasp in a synthetic way the effects of the change in economic structure, and in particular we focused on the changing allocation of labour force across sectors.

Table 9.2 Results of "reduced-form" VAR estimation of Equation (9.18)

		(1)	(2)	(3)	(4)	(5)	(6)	(7)	(8)	(9)
	VARIABLES	μ	π	α	*Koh*	*CD*	*Kap*	*TV*	*RTV*	*UTV*
β_{t-1}	Koh	–0.012	0.000	0.357***	–0.887***	0.163***	–0.116	–0.052	–0.327***	0.048
		–0.008	–0.002	–0.082	–0.017	–0.050	–0.115	–0.100	–0.117	–0.146
	Kap	–0.002	0.000	0.0343**	0.00798***	–0.014	–0.549***	0.009	0.0493**	–0.006
		–0.002	0.000	–0.015	–0.003	–0.009	–0.022	–0.018	–0.022	–0.028
	CD	0.001	–0.000462*	0.004	–0.00577**	–1.198***	0.005	0.0379**	0.0442**	–0.0487**
		–0.001	0.000	–0.014	–0.003	–0.008	–0.017	–0.016	–0.019	–0.024
	TV	0.001	–0.002	0.064	0.019	–0.570***	0.275	–0.369**	0.277	0.408**
		–0.012	–0.002	–0.121	–0.024	–0.073	–0.170	–0.146	–0.169	–0.207
	RTV	–0.001	0.001	–0.051	–0.005	0.329***	–0.167	0.007	–0.603***	–0.256**
		–0.007	–0.001	–0.073	–0.015	–0.044	–0.103	–0.089	–0.102	–0.124
	UTV	0.000	0.001	0.000	0.000	0.175***	–0.141***	0.009	–0.007	–0.656***
		–0.004	–0.001	–0.037	–0.007	–0.023	–0.054	–0.046	–0.051	–0.066
	μ	–0.618***	–0.001	–0.494	–0.091	–0.062	–0.052	0.013	–0.114	0.134
		–0.031	–0.006	–0.324	–0.062	–0.187	–0.430	–0.366	–0.447	–0.552
	π	–0.548***	–0.691***	–1.222	0.321	–0.248	–3.765*	1.115	2.074	5.892**
		–0.147	–0.029	–1.523	–0.289	–0.877	–2.071	–1.736	–2.023	–2.558
	α	0.00492**	0.00156***	–0.724***	–0.00930*	–0.0435***	0.043	0.038	0.010	0.059
		–0.002	0.000	–0.026	–0.005	–0.015	–0.033	–0.029	–0.035	–0.044
β_{t-2}	Koh	–0.001	0.00328**	0.101	–0.526***	0.106*	–0.091	0.226**	–0.123	0.629***
		–0.008	–0.001	–0.078	–0.019	–0.055	–0.120	–0.110	–0.132	–0.156
	Kap	–0.001	0.000	0.020	0.00858***	0.000	–0.382***	–0.011	0.033	–0.010
		–0.002	0.000	–0.016	–0.003	–0.010	–0.023	–0.019	–0.024	–0.029
	CD	0.001	–0.000781**	–0.001	–0.00593*	–0.823***	0.002	0.026	0.025	–0.0740**
		–0.002	0.000	–0.017	–0.003	–0.010	–0.023	–0.019	–0.024	–0.030
	TV	0.003	–0.00606**	–0.003	–0.0950***	0.137	0.817***	–0.028	0.110	0.188
		–0.013	–0.003	–0.132	–0.027	–0.084	–0.187	–0.165	–0.189	–0.230
	RTV	–0.001	0.00324**	–0.035	0.0707***	–0.062	–0.521***	–0.076	–0.321***	–0.078
		–0.008	–0.002	–0.079	–0.016	–0.051	–0.113	–0.100	–0.114	–0.136
	UTV	0.001	0.00164**	0.010	0.0258***	–0.016	–0.315***	–0.035	0.024	–0.247***
		–0.004	–0.001	–0.038	–0.008	–0.026	–0.056	–0.051	–0.056	–0.072

β_{t-2}	μ	–0.327***	–0.0210***	0.403	0.019	–0.305	0.364	–0.561	–0.554	0.752
		–0.034	–0.007	–0.354	–0.068	–0.205	–0.476	–0.400	–0.492	–0.601
	π	0.476***	–0.469***	–1.884	0.979***	–3.110***	–5.204**	–3.913*	–1.289	0.568
		–0.172	–0.035	–1.805	–0.355	–1.062	–2.463	–2.090	–2.529	–3.138
	α	0.003	0.00171***	–0.543***	–0.0110**	–0.0321**	0.027	0.050	0.003	0.0781*
		–0.003	–0.001	–0.028	–0.005	–0.016	–0.037	–0.032	–0.038	–0.047
β_{t-3}	Koh	–0.00952*	0.00217***	0.048	–0.213***	0.011	0.015	0.105	–0.200**	0.339***
		–0.005	–0.001	–0.052	–0.014	–0.040	–0.089	–0.082	–0.098	–0.116
	Kap	–0.002	0.000	0.019	–0.001	0.003	–0.0918***	–0.021	–0.018	–0.024
		–0.001	0.000	–0.014	–0.003	–0.008	–0.019	–0.016	–0.019	–0.025
	CD	–0.001	–0.000581**	0.0346**	–0.003	–0.399***	–0.004	0.0523***	0.0823***	–0.0504**
		–0.001	0.000	–0.014	–0.003	–0.008	–0.018	–0.015	–0.019	–0.023
	TV	0.004	–0.002	0.334***	–0.0770***	0.183**	0.663***	–0.865***	–1.090***	–0.015
		–0.011	–0.002	–0.119	–0.024	–0.073	–0.164	–0.144	–0.175	–0.213
	RTV	0.003	0.000	–0.221***	0.0671***	–0.116***	–0.439***	0.347***	0.316***	0.031
		–0.006	–0.001	–0.072	–0.014	–0.044	–0.099	–0.087	–0.105	–0.127
	UTV	0.002	0.000	–0.107***	0.0272***	–0.023	–0.207***	0.144***	0.320***	–0.309***
		–0.003	–0.001	–0.036	–0.007	–0.023	–0.048	–0.044	–0.054	–0.065
	μ	–0.0970***	–0.0161***	–0.099	0.023	–0.222	0.145	–0.617*	–0.788*	0.253
		–0.028	–0.006	–0.291	–0.056	–0.169	–0.387	–0.335	–0.404	–0.497
	π	0.382**	–0.292***	2.362	0.092	–4.403***	–2.540	–3.320*	–4.620**	2.027
		–0.151	–0.031	–1.570	–0.305	–0.931	–2.126	–1.829	–2.183	–2.738
	α	0.0113***	0.00120***	–0.245***	–0.001	–0.014	–0.003	–0.001	–0.008	0.015
		–0.002	0.000	–0.021	–0.004	–0.012	–0.028	–0.024	–0.028	–0.035
	Constant	0.000	0.000218***	–0.00323**	0.00184***	0.001	0.001	0.000	0.002	–0.002
		0.000	0.000	–0.001	0.000	–0.001	–0.002	–0.002	–0.002	–0.002
	R^2	0.178	0.209	0.265	0.301	0.378	0.170	0.195	0.204	0.225
	Observations	935	935	935	926	926	922	926	926	926

The empirical analysis, given the dynamic effects feeding back from economic and knowledge structure and vice versa, has been conducted by implementing a set of "reduced-form" VARs, which allowed us to investigate the lead–lag relationships between the two systems without imposing any a priori causal structure.

The results of the analysis are encouraging and call for further research in this direction, showing clear interactive patterns between the two structures. Changes in knowledge structure that signal the undertaking of exploitation strategies based on organized screening are likely to engender increasing within-sector productivity growth, while exploration strategies are likely to be followed by the changing allocation of labour force across sectors. We also noted how the increasing share of faster growing-sectors stimulates the establishment of exploitation activities dominated by organized search strategies within the comfortable boundaries of established competences. Moreover, the implementation of VAR(3) allowed us to appreciate also some interesting dynamics, like that relating variety and μ, which can be according to which increasing variety fosters the changing allocation of labour across sectors, but that this in turn is likely to be followed by a reduction in variety.

These results are obviously somewhat preliminary and do not pretend to have a final word on the relationship between knowledge and economic structure. We think, however, that they are interesting both with respect to the mechanisms on which they shed new light and with respect to the identification of new methodological approaches to address these issues.

Notes

1 We acknowledge that the use of administrative regions to investigate the effects of knowledge creation represents only an approximation of the local dynamics underpinning such process. Indeed administrative borders are arbitrary, and therefore might not be representative of the spontaneous emergence of local interactions. It would be much better to investigate these dynamics by focusing on local systems of innovation. However, it is impossible to find data at such a level of aggregation. Moreover, the identification of local systems involve the choice of indicators and threshold values according to which one can decide whether to unbundle or not local institutions. This choice is in turn arbitrary, and therefore it would not solve the problem, but it would only reproduce the issue at a different level. Thus we think that despite the unavoidable approximation, our analysis may provide useful information on the dynamics under scrutiny.

2 The assignment of patent to regions on the basis of inventors' addresses is the most widespread practice in the literature (see, for example, Maurseth and Verspagen, 2002; Henderson *et al.*, 2005; Breschi and Lissoni, 2009, Paci and Usai, 2009, to cite a few). A viable alternative may rest on the use of applicants' addresses, above all when the assessment of knowledge impact on growth is at stake (see Antonelli *et al.*, 2010). However, when the analysis is conducted at the local level of aggregation, and the geography of collective processes of knowledge creation is emphasized, the choice of inventors' addresses remains the best one.

10 Conclusions

This book has elaborated the idea that knowledge is an evolving complex system. More specifically, we contend that knowledge is a subsystem which is part of a hierarchy of nested subsystems characterized by a recursive structure. Building upon the theory of complex-system dynamics, we have characterized the structure of knowledge as an outcome of a combinatorial process, according to which can safely be represented a network the nodes of which are concepts, small units of knowledge, and the links the actual combinations among such concepts. New knowledge emerges out of dynamics of combination implemented through search strategies which are conducted across the knowledge landscape. Traditional concepts like preferential attachment and fitness, pleiotropy and epistatic relationships, all apply to the analysis of knowledge structure, and in particular to the analysis of the change in the structure of knowledge. The title of the book is exactly meant to emphasize the dynamic dimension of knowledge structure, that is likely to change endogenously as an effect of feedbacks and interactions with the other subsystems, and in particular with the innovation and the economic ones.

We decided to build the idea of a changing structure of knowledge by first elaborating on the notion of structural change as it is used in economics. The revival of structuralism in economics can be considered in this respect a second-order purpose of this book. The analysis of structural change indeed has been neglected for much a long time by scholars too preoccupied by the search for a stable equilibrium to allow growth rates to be unevenly distributed across nations and sectors.

The analysis of structural change in economics has such venerable origins that its underdevelopment really seems unacceptable. In Chapter 1 we have provided the empirical motivation to an analysis of structural change both in the traditional sense and applied to knowledge viewed as an organized structure. The empirical evidence of the last decades indeed speaks for an increasing change in employment patterns in most advanced countries, along with the establishment of new technological paradigms. Changes in the knowledge bases of advanced countries have indeed led to that powerful convergence of different and yet related technologies which we label today information and communication technologies, or, more familiarly, ICTs. The widespread diffusion of such technologies has favoured, and it has been favoured by, the changing employment specialization characterized by decreasing shares of manufacturing sectors and increasing shares of dedicated business services. Another important enabling condition in the structure of labour markets has been represented by the entry on the supply side of highly skilled workers, endowed with competences more and more developed through the access to formal education institutions. The structure of labour markets interact therefore with the structure and that of technological knowledge. The sets of mutual interdependences among these systems, as well as among them and the rest of the socio-economic system, is quite difficult to be commanded at once, a feature which will be sufficient to label such interaction as complex à la Hayek.

Structural change appears therefore to be an inherent feature of economic systems, as well as of each of the other interconnected subsystems. After all, the former implicit treatments of the subject can be traced as back far as the seminal Adam Smith work, *Wealth of Nations*. Simon Kuznets is traditionally acknowledged as the founding father of the analysis of the changing structure of the economy although, as we have argued in Chapter 2, an important and insightful antecedent is Marshall's *Industry and Trade*, eventually complemented by the analysis put forward in the *Principles* (Marshall, 1890, 1919). Marshall's arguments are still at the basis of the actual investigation of the consequences of the interplay between industrial specialization and cross-country uneven development of sectors. His line of reasoning can well be articulated using the lexicon of complex systems. Variations in the compositions of imports and exports are likely to reflect the changing patterns of fitness values of sectors within a particular context. Fitness values are in turn shaped by technological improvement, which is in turn an emergent property of knowledge, innovation and productive systems. The evolution of knowledge once again is maintained to be a crucial element for the change in the sectoral composition of economic systems. The combination of Marshall's arguments with Smith's considerations on division of labour has allowed Young (1928) to propose a dynamic representation of structural change driven by the increasing scale of production. In this perspective, structural change is generated not only by the changing weight of sectors in the economy, but also by the creation of brand new sectors. In a complex-system approach the sectors constituting the economic structure could be thought of as modules featuring mutual interdependence. This is exactly the representation given by Young, who emphasized the dynamic aspect of horizontal and vertical relationships. Even more, Young theorized the existence of generative relationships according to which the interactions among modules of a subsystems generate new modules, giving rise to an endless process of change.

As already said, Kuznets was the first scholar to have attempted a systematic analysis of the process of structural change. His work has been very influential, above all regarding the implications in terms of economic convergence, a topic much *à la mode* in the late 1980s and early 1990s. An important missing link in his body of work remains, however, the proper appreciation of the role of innovation and technological knowledge. This makes Kuznets' contributions quite complementary to those of Schumpeter, who in turn failed to provide a proper account of the effects and consequences of innovation on structural change. The intertwining of the two streams of literature is therefore a crucial step to the appreciation of the dynamic interactions between the economic and the knowledge structure.

The application of complex-systems theory to the articulation of an integrated framework to the analysis of knowledge structure as a module of a wider system, of a nested hierarchy, has proved to provide a useful heuristics to investigate the effects of changes of knowledge structure on different aspects of the economy, at different levels of aggregation. We have proposed building our framework on the unifying concept of economic space as proposed by François Perroux. Space in this perspective is not meant to refer to geography. Such concept has a relational definition, and identifies a bundle of forces emanating from the elements in the sets of relations. The idea of space and that of structure overlaps also in complexity scholars, who operationalize the behaviour of agents by using the metaphor of landscapes. The revival of structuralism which we have proposed allows us to go beyond a relevant limit of most scholars dealing with complexity, according to which the structure of complex systems is made of a stable number of components. In this perspective one would be able only to understand what is a particular design of a complex system, without understanding how it can be changed. The approach of genetic structuralism allows for introducing history, and hence evolution, into the analysis of complex systems. In view of this, emergent properties arise not only from dynamic interactions, but also from changes in the structure of interactions, say the introduction of new components or the change in the architecture of the relations, which are generated by dynamic relations themselves. Generative relationships are therefore likely to affect the structure of the system, and eventually the outcome of dynamic interactions. From an economic viewpoint, the structure of knowledge, which is an emergent property generated mainly by dynamic interactions in the innovation system, can change endogenously, i.e. as an effect of forces that are internal to the economic system. Agents' behaviour, be they consumers, firms, researchers, is likely to identify the conditions to put forth new knowledge by means of new combination possibilities among existing concepts, or by introducing new concepts. In this respect the distinction may be useful between new ideas that are new in that they are generated from other fields, and ideas that are new in that they did not exist at all before. This latter event is much more rare, and is likely to generate a sharp discontinuity in the technological evolution. The former is more likely to enhance generative creativity, and makes more desirable a brokerage than a cohesive structure (Fleming *et al.*, 2007).

The endogenous modification of knowledge structure in turn shapes the changes in the economic structure, and is also likely to go with changes in the institutional and the education systems. These in turn generate other changes which will be reflected in the rest of the system. As a result, such nested hierarchy can hardly reach a stable equilibrium. On the whole we are instead confronted with a restless process of change, in which dynamic interactions are always in motion and generate further interactions.

Besides the theoretical aspects, the conceptualization of knowledge derived in Chapter 4 turned out to be useful in overcoming the traditional operational translation of knowledge in empirical analyses. Beyond the concept of knowledge capital stock, we have shown in Chapter 5 that the network representation of knowledge is so flexible that it allows us to derive different indicators by using different methodological approaches. The utilization of co-occurrence matrices has proved to be useful to derive statistical indicators like coherence, cognitive distance or variety. We have emphasized the similarities between the co-occurrence matrices and the design structure matrices used in the complexity-based analysis of product technologies. In addition, we have noticed how the implementation of social network analysis can be far-reaching in providing a way to investigate changes of knowledge structure, as well as the relationships between changes in the structure of knowledge and changes in the structure of coalitions for innovation. The last part of the book has provided some examples of the flexibilities of such methodologies, showing that they are suitable to describe the structure of the knowledge base at different levels of aggregation, say regions, countries or sectors.

The adoption of such theoretical perspective bears important consequences also regarding the design of technology policies. Innovation policies have indeed increasingly become the strategic lever aimed at rejuvenating the growth process in mature industries and at creating the conditions for the birth of new industries, as well as at providing the means by which less developed areas could reduce the gaps with advanced economies. However, innovation policies have mainly been designed to work on the supply side of the knowledge production process. The traditional approach to innovation policies places in the correction of "market failures" the main rationale for public intervention. Based on the analysis by Kenneth Arrow (1962), knowledge is basically considered to be information, and hence described as a public good characterized by non-rivalry, non-excludability and non-appropriability. In this direction, most innovation policies involved the dissemination of incentives and subsidies to compensate would-be innovators for the pretended non-appropriability of knowledge and hence remedy to ensuing market failures.

The evolutionary perspective on science and technology policy represents a clear step forward (Lundvall, 1992; Nelson, 1993; Edquist, 1997). First of all, the view of knowledge as a public good is abandoned to propose a more realistic view of knowledge as embodied in the idiosyncratic features of economic agents and of the networks within which they conduct their research. In this framework knowledge is essentially tacit, and much emphasis is put on learning dynamics and skills development, as well as on the interactive nature of the innovation process occurring within innovation systems defined at both national and regional level (Rosenberg, 1990; Pavitt, 1998). As Salter and Martin (2001) show, the evolutionary approach to technology policy allows for the appreciation of a wider set of economic and social benefits, that go well beyond the simplistic argument concerning market failures. They identify six classes of benefits that may accrue from innovation policies, which relate to (1) the increase in the stock of available knowledge;

(2) training of skilled graduates; (3) creating new scientific instrumentation and methodologies; (4) networks and social interactions; (5) problem-solving; (6) creation of new firms. Although grounded on a more advanced understanding of the innovation process, these policies share with the evolutionary and the innovation system approaches a pretty deterministic view on technologies, which are supposed to follow defined stages of a lifecycle once introduced, and the idea that systemic ties are given by nature and only wait to be fuelled. Moreover, such policies are also mostly oriented to the supply-side of knowledge generation.

The approach developed in this book allows us to appreciate the dynamic interactions among different subsystems as well as within subsystems themselves. The mutually interdependent nature of these systems is such that each change in one part of the whole is likely to affect many other parts, depending on the intensity of their relationships according to the concept of pleiotropy. Moreover, dynamic interactions are likely to generate new modules and therefore new links as an effect of exaptive bootstrap and generative relationships. In other words, the main message of this book is that *structural change is endogenous*. The structures of the subsystems change as an effect of dynamic interactions occurring at the agents' layer, which are in turn shaped by architectures that they themselves contribute to modify. In this perspective, any approach to technology policy could not be anything but systemic. However, the policy design should take into account the fact that structures are neither stable nor exogenous. After all, policymakers are themselves part of the agents' layer. In this perspective policy measures can be regarded as emergent properties stemming from the institutional subsystem. However, these should emerge through a complex set of dynamic feedbacks with the other subsystems. The target of a technology in this perspective goes far beyond the providing of incentives in a generic way. They could also be directed to foster changes in the structure of interactions among the agents within the innovation system. They could also aim at facilitating the entry of new kinds of agent within the network of innovating agents. A clear example in this direction is the increasing role of venture capitalists in the financing of innovations. The set of norms that made this possible has favoured the inclusion of new agents in the financial subsystem, which clearly overlaps with the innovation one. A systemic approach also allows us to better appreciate the opportunities to act on other subsystems that are likely to exert an influence on knowledge interactions. In particular, most traditional approaches move from the idea that knowledge producing agents should receive incentives as the results of their interactions are likely to positively affect economic performances. However, the relationships also go the other way. The economic performances are likely to shape technology dynamics. The complexity-based approach to knowledge and economic structure allows us therefore to reconcile supply- and demand-side policy measures in a unified framework. An important issue in this respect concerns the identification of the relevant parts of the economic system that are likely to most influence the knowledge system. Theorists of the active role of users (von Hippel, 1988; Rosenberg, 1982) would suggest that demand matters not only because it provides the money to bear investments in the creation of new knowledge, but it is also the main source of knowledge inputs in such process. The classical demand–pull approach, which at least in its modern form was undoubtedly pioneered by Jacob Schmookler, would instead emphasize the importance of the availability of resources to commit to knowledge creation. He observed how series on technology creation as proxied by patent applications tend to follow series on output (Schmookler, 1954, 1962). The suggested interpretation of this evidence was grounded on the idea that

> more money will be available for invention when the industry's sales are high than when they are low. Increased sales imply that both the producing firms and their employees will be in a better position than before to bear the expenses of invention.
>
> (Schmookler, 1962, p. 17)

In this framework, the ability to finance the activities of knowledge creation plays a central role (Schmookler, 1966).

The appreciation of dynamic interactions between the knowledge and economic systems should therefore push policymakers to reconsider the possibility to implement *Keynesian schemes of intervention to spur knowledge creation*. From a theoretical viewpoint, this would amount to going beyond the limits of the Keynesian framework by putting the discourse on the role of public procurement into a dynamic framework. Schumpeter (1936, 1946) indeed emphasized how Keynes' *General Theory* could have been regarded as an example of "macrostastics" rather than macrodynamics, as his argument was based on the consideration that variations in output are solely related to variations in employment, neglecting therefore the major aspect of capitalist economies, i.e. the change in productive capacities and production techniques. In this respect, public procurement can be designed instead as an instrument to promote knowledge creation by increasing the demand of strategic activities. Moreover, according to Schumpeter's statement on efficiency in government expenditure programmes, public demand risks being ineffective, like a generic drug used to treat a specific disease. On the contrary, government intervention should be carefully designed so as to be targeted towards the most promising elements of the system. This requires the capacity to explore the knowledge and industrial landscape in order to identify the areas deserving the government support. The architecture of subsystems, as well as of inter-systemic connections, can be very informative in this respect. Technology policy should be addressed towards key points in the architecture, the improvement of which is more likely to affect the rest of the system in a pretty significant way. Pleiotropy is therefore a key feature to take into account. It should be coupled with considerations about fitness values of the domains candidate to receive government funding. Most profitable technologies, or industries in which the phase of random search has been superseded by organized search, can represent good targets. Once again, the architecture of knowledge structure interacts with another part of the system, in this case the institutional subsystem, in the generation of an emergent property.

The case for a systemic approach to policy in general, and technology policy in particular, emerges. The potential for improving the effectiveness of state intervention can be far-reaching. Obviously we have provided here only some hints of how pervasive the proper account of the endogenous nature of structural change in knowledge and economy can be. Complex socio-economic systems are after all populated by interacting agents who ultimately are living persons, and life, as Pirandello explained, does not conclude. In the same vein, this book does not conclude, and we can only manifest the need for further research within such a fascinating framework to the understanding of the dynamics of human actions and of their outcome.

References

Abramovitz, M., 1956. Resource and output trends in the United States since 1870. *American Economic Review*, 46, 5–23.

Abramovitz, M., 1986. Catching up, forging ahead, and falling behind. *Journal of Economic History*, 46, 2, 385–406.

Abramovitz, M., 1994. The origins of the postwar catch-up and convergence boom, in J. Fagerberg, B. Verspagen and N. von Tunzelmann (eds), *The Dynamics of Technology, Trade and Growth*. Aldershot: Edward Elgar, pp. 21–52.

Acemoglu, D. and Guerrieri, V., 2008. Capital deepening and non-balanced economic growth. *Journal of Political Economy*, 116, 3, 467–498.

Acs, Z., Anselin, L., and Varga, A., 2002. Patents and innovation counts as measures of regional production of new knowledge. *Research Policy*, 31, 1069–1085.

Agarwal, R. and Tripsas, M., 2008. Technology and industry evolution, in S. Shane (ed.), *Handbook of Technology and Innovation Management*. New York: Wiley, pp. 3–55.

Aghion, P. and Howitt, P., 1992. A model of growth through creative destruction. *Econometrica*, 60, 323–351.

Agrawal, A., Cockburn, I. and McHale, J., 2006. Gone but not forgotten: knowledge flows, labor mobility, and enduring social relationships. *Journal of Economic Geography*, 6, 571–591.

Allen, R.C., 1983. Collective invention. *Journal of Economic Behavior and Organization*, 4, 1–24.

Altenberg, L., 1994. Evolving better representations through selective genome growth, in J.D. Schaffer, H.P. Schwefel and H. Kitano (eds), *Proceedings of the IEEE World Congress on Computational Intelligence*. Piscataway, NJ: IEEE, pp. 182–187.

Altenberg, L., 1995. Genome growth and the evolution of the genotype–phenotype map, in W. Banzhaf and F.H. Eckman (eds), *Evolution and Biocomputation*. Berlin and Heidelberg: Springer-Verlag, pp. 205–259.

Amesse, F. and Cohendet, P., 2001. Technology transfer revisited from the perspective of the knowledge-based economy. *Research Policy*, 1459–1478.

Anselin, L., 1988. *Spatial Econometrics: Methods and Models*. Dordrecht: Kluwer.

Antonelli, C., 1999. *The Microdynamics of Technological Change*. London: Routledge.

Antonelli, C., 2001. *The Microeconomics of Technological Systems*. Oxford: Oxford University Press.

Antonelli, C., 2003. *The Economics of Innovation, New Technologies and Structural Change*. London: Routledge.

Antonelli, C., 2005. Models of knowledge and systems of governance. *Journal of Institutional Economics*, 1, 51–74.

Antonelli, C., 2008. *Localized Technological Change: Towards the Economics of Complexity*. London: Routledge.

Antonelli, C., and Calderini, M., 2008. The governance of knowledge compositeness and technological performance: the case of the automotive industry in Europe. *Economics of Innovation and New Technology*, 17, 23–41.

Antonelli, C., Krafft, J. and Quatraro, F., 2010. Recombinant knowledge and growth: the case of ICTs. *Structural Change and Economic Dynamics*, 21, 50–69.

Antonelli, C., Patrucco, P.P. and Quatraro, F., 2007. The economics of new information and communication technologies, in W. Donsbach (ed.), *The International Encyclopedia of Communication*. Oxford: Blackwell.

Antonelli, C., Patrucco, P.P. and Quatraro, F., 2008. The economics of information and communication technologies, in W. Donsbach (ed.), *The Encyclopaedia of Communication*. Oxford: Blackwell, pp. 2225–2230.

Antonelli, C., Patrucco, P.P. and Quatraro, F., 2011. Pecuniary knowledge externalities: evidence from European regions. *Economic Geography*, 87, 23–50.

Arrow, K.J., 1962. Economic welfare and the allocation of resources for invention, in R.R. Nelson (ed.), *The Rate and Direction of Inventive Activity: Economic and Social Factors*. Princeton, NJ: Princeton University Press for NBER, pp. 609–625.

Arora, A. and Gambardella, A., 1994. The changing technology of technological change: general and abstract knowledge and the division of innovative labour. *Research Policy*, 23, 523–532.

Arora, A., Fosfuri, A. and Gambardella, A., 2001. *Markets for Technology: The Economics of Innovation and Corporate Strategy*. Cambridge, MA: MIT Press.

Arthur, W.B., 2009. *The Nature of Technology*. New York: Free Press.

Bain, J.S., 1956. *Barriers to New Competition*. Cambridge: Cambridge University Press.

Balconi, M., Brusoni, S. and Orsenigo, L., 2009. In defense of the linear model: an essay, *Research Policy*.

Baldwin, C.Y., 2007. Where do transactions come from? Modularity, transactions, and the boundaries of firms. *Industrial and Corporate Change*, 17, 155–195.

Baldwin, C.Y. and Clark, K.B., 2000. *Design Rules*, Vol. I: *The Power of Modernity*. Cambridge, MA: MIT Press.

Barabasi, A.L. and Albert, R., 1999. Emergence of scaling in random networks. *Science*, 286, 509–512.

Barabasi, A.L., Jeong, H., Neda, Z., Ravasz, E., Schubert, A. and Vicsek, T., 2002. Evolution of the social network of scientific collaborations, *Physica A*, 311, 590–614.

Bathelt, H., Malmberg, A. and Maskell, P., 2004. Clusters and knowledge: local buzz, global pipelines and the process of knowledge creation. *Progress in Human Geography*, 28, 31–56.

Baumol, W.J., 1967. Macroeconomics of unbalanced growth: the anatomy of urban crisis. *American Economic Review*, 57, 415–426.

Baumol, W., 1986. Productivity growth, convergence and welfare: what the long-run data show. *American Economic Review*, 76, 1072–1085.

Baumol, W.J., Blackman, S.A.B. and Wolff, E.N., 1985. Unbalanced growth revisited: asymptotic stagnancy and new evidence. *American Economic Review*, 75, 806–817.

Baumol, W.J., Blackman, S.A.B. and Wolff, E.N., 1989. *Productivity and American Leadership: The Long View*. Cambridge, MA: MIT Press.

Beugelsdijk, S., 2007. The regional environment and a firm's innovative performance: a plea for a multilevel interactionist approach. *Economic Geography*, 83, 181–199.

Bianconi, G. and Barabasi, A.L., 2001. Bose-Einstein condensation in complex networks. *Physical Review Letters*, 86, 5632–5635.

Binder, M., Hsiao, C. and Pesaran, C.H., 2005. Estimation and inference in short panel vector autoregressions with unit roots and cointegration. *Econometric Theory*, 21, 795–837.

Bloch, H. and Metcalfe, J.S., 2011. Complexity in the theory of the developing firm, in C. Antonelli (ed.), *Handbook on the Economic Complexity of Technological Change*. Cheltenham: Edward Elgar.

Blundell, R.W. and Bond, S.R., 1998. Initial conditions and moment restrictions in dynamic panel data models. *Journal of Econometrics*, 87, 115–143.

Boschma, R., 2005. Proximity and innovation: a critical assessment. *Regional Studies*, 39, 61–74.

Botazzi, G. and Secchi, A., 2006. Explaining the distribution of firms' growth rates. *Rand Journal of Economics*, 37, 234–263.

Botazzi,, G., Cefis, E., Dosi, G. and Secchi, A., 2007. Invariances and diversities in the patterns of industrial evolution: some evidence from Italian manufacturing industries. *Small Business Economics*, 29, 137–159.

Bouraoui, A., 2009. Hayek, l'"ordre spontané" et la complexité. *Revue Economique*, 60, 1355–1358.

Breschi, S. and Lissoni, F., 2001. Knowledge spillovers and local innovation systems: a critical survey. *Industrial and Corporate Change*, 10, 975–1005.

Breschi, S. and Lissoni, F., 2009. Mobility of skilled workers and co-invention networks: an anatomy of localized knowledge flows. *Journal of Economic Geography*, 94, 439–468.

Breschi, S., Lissoni, F. and Malerba, F., 2003. Knowledge relatedness in firm technological diversification. *Research Policy*, 32, 69–97.

Bresnahan, T.F. and Malerba, F., 1999. Industrial dynamics and the evolution of firms' and nations' competitive capabilities in world computer industry, in D.C. Mowery and R.R. Nelson (eds), *Sources of Industrial Leadership: Studies of Seven Industries*. Cambridge: Cambridge University Press, pp. 79–133.

Brower, E. and Kleinknecht, A., 1999. Innovative output and a firm's propensity to patent: an exploration of CIS micro data. *Research Policy*, 28, 615–624.

Buchanan, J.M., 1965. An economic theory of clubs. *Economica*, 34, 1–14.

Buerger, M., Broekel, T. and Coad, A., 2011. Regional dynamics of innovation: investigating the co-evolution of patents, research and development (R&D), and employment. *Regional Studies*, forthcoming.

Burns, A.F., 1934. *Production Trends in the United States since 1870*. Boston: NBER.

Bush, V., 1945. *Science the Endless Frontier: Report to the President*, Washington, DC: United States Government Printing Office.

Castaldi, C. and Dosi, G., 2009. The patterns of output growth of firms and countries: scale invariances and scale specificities. *Empirical Economics*, 37, 475–495.

Chenery, H.B., 1960. Patterns of industrial growth. *American Economic Review*, 50, 624–654.

Coad, A., 2010. Exploring the processes of firm growth: evidence from a vector autoregression. *Industrial and Corporate Change*, 19, 1677–1703.

Coe, D. and Helpman, E., 1995. International R&D spillovers. *European Economic Review*, 39, 859–887.

Cohen, W.M. and Levinthal, D.A., 1990. Absorptive capacity: a new perspective on learning and innovation. *Administrative Science Quarterly*, 35, 128–152.

Colombelli, A. and von Tunzelmann, G.N., 2011. The persistence of innovation and path dependence, in C. Antonelli (ed.), *Handbook on the Economic Complexity of Technological Change*. Cheltenham: Edward Elgar.

Colombelli, A., Krafft, J. and Quatraro, F., 2011. High growth firms and technological knowledge: do gazelles follow exploration or exploitation strategies? Submitted to *Industrial and Corporate Change*.

Cooke, P., 1996. The new wave of regional innovation networks: analysis, characteristics and strategy. *Small Business Economics*, 8, 1–13.

Cooke, P., Uranga, M. and Etxebarria, G., 1997. Regional innovation systems: institutional and organizational dimensions. *Research Policy*, 26, 475–491.

Corning, P.A., 2002. The re-emergence of emergence: a venerable concept in search of a theory. *Complexity*, 7, 6, 18–30.

Corrocher, N., Malerba, F. and Montobbio, F., 2007. Schumpeterian patterns of innovative activity in the ICT field. *Research Policy*, 36, 418–432.

Corry, L., 1992. Nicolas Bourbaki and the concept of mathematical structure. *Synthese*, 92, 315–348.

Cowan, R. and Foray, D., 1997. The economics of codification and the diffusion of knowledge. *Industrial and Corporate Change*, 6, 3, 595–622.

Cowan, R. and Jonard, R., 2003. The dynamics of collective invention. *Journal of Economic Behavior and Organization*, 52, 513–532.

Cowan, R., David, P. and Foray, D., 2000. The Explicit economics of knowledge codification and tacitness. *Industrial and Corporate Change*, 9, 211–253.

Creamer, D., 1942. Shifts of manufacturing industries, in *Industrial Location and National Resources*. Washington, DC: U.S. National Resources Planning Board.

Crescenzi, R., Rodriguez-Pose, A. and Storper, M., 2007. The territorial dynamics of innovation: a Europe–United States comparative analysis. *Journal of Economic Geography*, 673–709.

Cuneo, P. and Mairesse, J., 1984. Productivity and R&D at the firm level in French manufacturing, in Z. Griliches (ed.), *R&D, Patents and Productivity*. Chicago: University of Chicago Press.

Daveri, F., 2002. The new economy in Europe, 1992–2001. *Oxford Review of Economic Policy*, 18, 345–362.

David., P., 2001. Understanding digital technology's evolution and the path of measured productivity growth: present and future in the mirror of the past, in E. Brynjolfsson and B. Kahin (eds), *Understanding Digital Economy*, Cambridge, MA: MIT Press.

D'Este, P., 2005. How do firms' knowledge bases affect intra-industry heterogeneity? An analysis of the Spanish pharmaceutical industry, *Research Policy*, 34, 1, 33–45.

De Kerchove, D., 1997. *Planetary Mind: Collective Intelligence in the Digital Age*. New York: Hardwired.

De Saussure, F., 1916. *Course de linguistique générale*, Paris: Payot.

Descombes, V., 1980. *Modern French Philosophy*. Cambridge: Cambridge University Press.

de Solla Price, D., 1965. Networks of scientific papers. *Science*, 149, 3683, 510–515.

Dicken, P. and Malmberg, A., 2001. Firms in territories: a relational perspective. *Economic Geography*, 77, 345–363.

Dosi, G., 1982. Technological paradigms and technological trajectories: a suggested interpretation of the determinants and directions of technical change. *Research Policy*, 11, 147–162.

Dosi, G., 1988. Sources, procedures, and microeconomic effects of innovation. *Journal of Economic Literature*, 26, 3, 1120–1171.

Dunning, J.H., 1993. *Alliance Capitalism and Global Business*. London: Routledge.
Echevarria, C., 1997. Changes in sectoral composition associated with economic growth. *International Economic Review*, 38, 431–452.
Edquist, C. (ed.), 1997. *Systems of Innovation: Technologies, Institutions and Organizations*. London: Pinter Publishers/Cassell Academic.
Elhorst, P.J., 2003. Specification and estimation of spatial panel data models. *International Regional Science Review*, 26, 244–268.
Englander, A.S., and Mittelstädt, 1988. Total factor productivity: macroeconomic and structural aspects of slowdown. *OECD Economic Studies*, 11.
Esteban, J.M., 1972. Shift-share analysis revisited. *Regional and Urban Economics*, 2, 249–261.
Esteban, J., 2000. Regional convergence in Europe and the industry mix: a shift-share analysis. *Regional Science and Urban Economics*, 30, 353–364.
Fagerberg, J., 1987. A technology gap approach to why growth rates differ. *Research Policy*, 16, 87–99.
Fagerberg, J., 1994. Technology and international differences in growth rates. *Journal of Economic Literature*, 32, 1147–1175.
Fagerberg, J., 2000. Technological progress, structural change and productivity growth: a comparative study. *Structural Change and Economic Dynamics*, 11, 393–411.
Feser, E.J., 2002. Tracing the sources of local external economies. *Urban Studies*, 39, 2485–2506.
Fisher, A.G.B., 1939. Production, primary, secondary and tertiary. *The Economic Record*, 15, 24–38.
Fleming, L., 2001. Recombinant uncertainty in technological search. *Management Science*, 47, 117–132.
Fleming, L. and Sorenson, O., 2001. Technology as a complex adaptive system: evidence from patent data. *Research Policy*, 30, 1019–1039.
Fleming, L., Mingo, S. and Chen, D., 2007. Collaborative brokerage, generative creativity and creative success. *Administrative Science Quarterly*, 52, 443–475.
Foray, D., 2004. *The Economics of Knowledge*. Cambridge, MA: MIT Press.
Fransman, M., 2002. *Telecoms in the Internet age: from boom to bust to…?* Oxford: Oxford University Press.
Fransman, M., 2007. *The New ICT Ecosystem: Implications for Europe*. Edinburgh: Kokoro.
Frantzen, D., 2000. Innovation, international technological diffusion and changing influence of R&D on productivity. *Cambridge Journal of Economics*, 24, 193–210.
Freeman, L.C., 1979. Centrality in social networks: conceptual clarification. *Social Networks*, 1, 215–239.
Freeman, C. and Soete, L., 1990. Fast structural change and slow productivity change: some paradoxes in the economics of information technology. *Structural Change and Economic Dynamics*, 1, 225–242.
Freeman, C., and Soete, L., 1997. *The Economics of Industrial Innovation*, 3rd edn. Cambridge, MA: MIT Press.
Frenken, K. and Nuvolari, A., 2004. Entropy statistics as a framework to analyse technological evolution, in J. Foster and W. Hölzl (eds), *Applied Evolutionary Economics and Complex Systems*, Cheltenham: Elgar.
Frenken, K., von Oort, F. and Verburg, T., 2007. Related variety, unrelated variety and regional economic growth. *Regional Studies*, 41, 685–697.
Fritsch, M., 2002. Measuring the quality of regional innovation systems: a knowledge production function approach. *International Review of Regional Science*, 25, 86–101.

Fritsch, M., 2004. Cooperation and the efficiency of regional R&D activities. *Cambridge Journal of Economics*, 28, 829–846.

Fritsch, M. and Franke, G., 2004. Innovation, regional knowledge spillovers and R&D cooperation. *Research Policy*, 33, 245–255.

Fuà, G., 1983. Industrializzazione nel Nord Est e nel Centro, in G. Fuà and C. Zacchia (eds), *Industrializzazione Senza Fratture*. Bologna: il Mulino, pp. 7–46.

Fuà, G. and Zacchia, C. (eds), 1983. *Industrializzazione Senza Fratture*, Bologna: il Mulino.

Garcia-Vega, M., 2006. Does technological diversification promote innovation? An empirical analysis for European firms. *Research Policy*, 35, 230–246.

Gibbons, M., Limoges, C., Nowtny, H., Schwartzman, S., Scott, P. and Trow, M., 1984. *The New Production of Knowledge: The Dynamics of Science and Research in Contemporary Societies*. London: Sage.

Giuliani, E., 2007. The selective nature of knowledge networks in clusters: evidence from the wine industry, *Journal of Economic Geography*, 7, 139–168.

Gossling, T., 2003. Proximity, trust and morality in networks. *European Planning Studies*, 12, 675–689.

Grabowski, H.G. and Vernon, J.M., 1994. Innovation and structural change in pharmaceuticals and biotechnology. *Industrial and Corporate Change*, 3, 435–449.

Griliches, Z., 1957. Hybrid corn: an exploration in the economics of technological change. *Econometrica*, 25, 4, 501–522; reprinted in Fox, K.A. and Johnson, D.G. (eds), 1969. *AEA Readings in Agricultural Economics*.

Griliches, Z., 1967. Distributed lags: a survey, *Econometrica*, 35, 16–49.

Griliches, Z., 1979. Issues in assessing the contribution of research and development to productivity growth. *The Bell Journal of Economics*, 10, 92–116.

Griliches, Z., 1980. R&D and the productivity slowdown. *American Economic Review*, 70, 343–348.

Griliches, Z. (ed.), 1984. *R&D, Patents and Productivity*. Chicago: University of Chicago Press.

Griliches, Z., 1990. Patent statistics as economic indicators: a survey. *Journal of Economic Literature*, 28, 1661–1707.

Griliches, Z., 1992. The search for R&D spillovers. *Scandinavian Journal of Economics*, 94, 29–47.

Grossman, G.M. and Helpman, E., 1991. *Innovation and Growth in the Global Economy*. Cambridge, MA: MIT Press.

Hall, B.H. and Ziedonis, R.H., 2001. The determinants of patenting in the US semiconductor industry, 1980–1994. *Rand Journal of Economics*, 32, 101–128.

Hall, B.H. and Ziedonis, R.H., 2007. An empirical analysis of patent litigation in the semiconductor industry. Working Paper, University of California, Berkeley, 31.

Hall, B.H., Griliches, Z. and Hausman, J.A., 1986. Patents and R&D: is there a lag? *International Economic Review*, 27, 265–283.

Hansen, B.E., 2001. The new econometrics of structural change: dating breaks in U.S. labor productivity. *Journal of Economic Perspectives*, 15, 117–128.

Harhoff, D. and Reitzig, M., 2004. Determinants of opposition against EPO patent grants: the case of biotechnology and pharmaceuticals. *International Journal of Industrial Organization*, 22, 443–480.

Harhoff, D., Scherer, F. and Vopel, K., 2003. Citations, family size, opposition and the value of patent rights. *Research Policy*, 32, 8, 1343–1363.

von Hayek, F.A., 1937. Economics and knowledge. *Economica*, 4, 33–54.

von Hayek, F., 1945. The use of knowledge in society. *American Economic Review*, 35, 519–530.
von Hayek, F.A., 1967. *Studies in Philosophy, Politics and Economics*. London: Routledge.
von Hayek, F., 1973. *Law, Legislation and Liberty*, French translation Audouin, R., 1980. *Droit, législation et liberté*, I: *Règles et Ordre*. Paris: PUF.
Henderson, R.M. and Clark, K.B., 1990. Architectural innovation: the reconfiguration of existing product technologies and the failure of established firms. *Administrative Science Quarterly*, 35, 9–30.
Henderson, R.M., Jaffe, A. and Trajtenberg, M., 2005. Patent citations and the geography of knowledge spillovers: a reassessment – comment. *American Economic Review*, 95, 416–464.
Hidalgo, C.A., 2009. The dynamics of economic complexity and the product space over a 42 year period. CID Working Paper No. 189, Harvard University.
Hidalgo, C.A., Klinger, B., Barabasi, A.L. and Hausman, R., 2007. The product space conditions the development of nations. *Science*, 317, 482–487.
Higon, D., 2007. The impact of R&D spillovers on UK manufacturing TFP: a dynamic panel approach. *Research Policy*, 36, 964–979.
Houston, G.B., 1967. The shift and share analysis of regional growth: a critique. *Southern Economic Journal*, 33, 577–581.
Hymes, D., 1972. Models of interaction of language and social life, in D. Hymes and J.J. Gumperz (eds), *The Ethnography of Communication*. New York: Holt, Rinehart and Winston.
Jaffe, A.B., 1986. Technological opportunity and spillovers of R&D: evidence from firms' patents, profits, and market value. *American Economic Review*, 76, 984–1001.
Jaffe, A., 1989. Real effects of academic research. *American Economic Review*, 79, 957–970.
Jaffe, A.B. and Trajtenberg, M., 2002. *Patents Citations and Innovations. A Window on the Knowledge Economy*. Cambridge, MA: MIT Press.
Jorgenson, D.W., 1995. *Productivity*, Vol. I: *Post-War US Economic Growth*. Cambridge, MA: MIT Press.
Jorgenson, D., 2001. Information technology and the U.S. economy. *American Economic Review*, 91, 1–32.
Jorgensen, D.W., Ho, M.S. and Stiroh, K.J., 2006. Potential growth of the US economy: will the productivity resurgence continue? *Business Economics*, 41, 7–16.
Katila, R. and Ahuja, G., 2002. Something old, something new: a longitudinal study of search behaviour and new product introduction. *Academy of Management Journal*, 45, 1183–1194.
Kaufman, S., 1993. *Origins of Order: Self-organization and Selection in Evolution*. Oxford: Oxford University Press.
Klepper, S., 1997. Industry life cycles. *Industrial and Corporate Change*, 6, 145–182.
Kline, S.J. and Rosenberg, N., 1986. An overview on innovation, in R. Landau and N. Rosenberg (eds), *The Positive Sum Strategy*. Washington, DC: National Academy Press.
Kongsamut, P., Rebelo, S. and Xie, D., 2001. Beyond balanced growth. *Review of Economic Studies*, 68, 869–882.
Krafft, J., 2004. Entry, exit and knowledge: evidence from a cluster in the info-communications industry, *Research Policy*, 33, 1687–1706.
Krafft, J., 2009. Profiting in the info-coms industry in the age of broadband: lessons and new considerations, *Technological Forecasting and Social Change*.

Krafft, J. and Quatraro, F., 2011. The dynamics of technological knowledge: from linearity to recombination, in C. Antonelli (ed.), *Handbook on the Economic Complexity of Technological Change*. Cheltenham: Edward Elgar, pp. 181–200.

Krafft, J., and Salies, E., 2008. The cost of switching Internet providers in the broadband industry, or why ADSL has diffused faster than other innovative technologies: evidence from the French case. *Research Policy*, 37, 706–719.

Krafft, J., Quatraro, F. and Saviotti, P.P., 2009. Evolution of the knowledge base in knowledge intensive sectors. LEI-BRICK Working Paper no. 06/2009.

Krafft, J., Quatraro, F. and Saviotti, P., 2011. The knowledge base evolution in biotechnology: A social network analysis. *Economics of Innovation and New Technology*, 20, 445–475.

Krueger, J.J., 2008. Productivity and structural change: a review of the literature. *Journal of Economic Surveys*, 22, 330–363.

Kuhn, T., 1962. *The Structure of Scientific Revolution*. Chicago: Chicago University Press.

Kuznets, S., 1930. *Secular Movements in Production and Prices*. Boston: Houghton Mifflin.

Kuznets, S., 1940. Schumpeter's business cycles. *American Economic Review*, 30, 257–271.

Kuznets, S., 1954. Underdeveloped countries and the pre-industrial phase in the advanced countries, in A.N. Agarwala and S.P. Singh (eds), *The Economics of Underdevelopment*. New York: Oxford University Press, pp. 135–153.

Kuznets, S., 1972. Innovation and adjustments in economic growth. *Swedish Journal of Economics*, 74, 431–451.

Kuznets, S., 1973. Modern economic growth: findings and reflections. *American Economic Review*, 63, 247–258.

Kuznets, S., 1977. Two centuries of economic growth: reflections on US experience. *American Economic Review*, 67, 1–14.

Kuznets, S., 1989. *Economic Development, the Family, and Income Distribution*. Cambridge: Cambridge University Press.

Laitner, J., 2000. Structural change and economic growth. *Review of Economic Studies*, 67, 545–561.

Lane, D.A., 2011. Complexity in the theory of the developing firm, in C. Antonelli (ed.), *Handbook on the Economic Complexity of Technological Change*. Cheltenham: Edward Elgar.

Lane, D.A., van Der Leeuw, S.E., Pumain, D. and West, G. (eds.), 2009. *Complexity Perspectives in Innovation and Social Change*. Berlin: Springer.

Laursen, K., 1998. Revealed comparative advantage and the alternatives as measures of international specialization. DRUID Working Paper no. 98–30.

Lawson, C. and Lorenz, E., 1999. Collective learning, tacit knowledge and regional innovative capacity. *Regional Studies*, 33, 305–317.

Leten, B., Belderbos, R. and Van Looy, B., 2007. Technological diversification, coherence and performance of firms. *Journal of Product Innovation Management*, 24, 567–577.

Le Sage, J.P., 1999. The theory and practice of spatial econometrics. Department of Economics, University of Toledo. Online: www.spatial-econometrics.org.

Levin, R., Klevorick, A., Nelson, R. and Winter, S., 1987. Appropriating the returns from industrial research and development. *Brookings Papers on Economic Activity*, 3, 783.

Levi-Strauss, C., 1958. *Anthropologie Structurale I*. Paris: Plon.

Levi-Strauss, C., 1962. *La pensée sauvage*. Paris: Plon.
Lichtenberg, F., 1992. R&D investment and international productivity differences. *NBER* Working Paper, no. 4161.
Loos, B. and Verspagen, B., 2000. R&D spillovers and productivity: evidence from U.S. manufacturing microdata. *Empirical Economics*, 25, 127–148.
Lotti, F. and Schivardi, F., 2005. Cross country differences in patent propensity: a firm-level investigation. *Giornale degli Economisti*, 64, 469–502.
Lundvall, B.A. (ed.), 1992. *National Systems of Innovation*. London: Pinters.
Lypsey, R.G., Carlaw, K.I. and Bekar, C.T., 2005. *Economic Transformations: General Purpose Technologies and Long Term Economic Growth*. New York: Oxford University Press.
McCraw, T., 2006. Schumpeter's business cycles as business history. *Business History Review*, 80, 231–261.
Machlup, F., 1963. *Essays in Economic Semantics*. Englewood Cliffs, NJ: Prentice-Hall.
Machlup, F., 1971. *The Production and Distribution of Knowledge in the United States*. Princeton, NJ: Princeton University Press.
McLuhan, M., 1964. *Understanding Media: The Extensions of Man*. New York: McGraw Hill.
McWilliams, A. and Smart, D.L., 1993. Efficiency v. structure-conduct-performance: implications for strategy research and practice. *Journal of Management*, 19, 63–78.
Maddison, A., 1987. Growth and slowdown in advanced capitalist economies: techniques of quantitative assessment. *Journal of Economic Literature*, 25, 649–698.
Maleki, E., 2000. Knowledge, competence and new technology. *Economics of Innovation and New Technology*, 9, 71–79.
Malerba, F. and Orsenigo, L., 1995. Schumpeterian patterns of innovation. *Cambridge Journal of Economics*, 19, 1, 47–65.
Mansfield, E., 1961. Technical change and the rate of imitation. *Econometrica*, 29, 4, 741–766.
Maffezzoli, M., 2006. Convergence across Italian regions and the role of technological catch up. *Topics in Macroeconomics*, 6, Article 15.
March, J., 1991. Exploration and exploitation in organizational learning. *Organization Science*, 2, 71–87.
Marshall, A., 1890 [1920]. *Principles of Economics*. London: Macmillan.
Marshall, A., 1919. *Industry and Trade. A Study of Industrial Technique and Business Organization; and of Their Influences on the Condition of Various Classes and Nations*. London: Macmillan.
Marx, K., 1885 [2003]. Schematic presentation of accumulation and supplementary remarks, in H. Hagemann, M. Landesmann and R. Scazzieri (eds), *The Economics of Structural Change*, Vol. I, *Critical Writings*. Aldershot: Edward Elgar.
Maurseth, P.B. and Verspagen, B., 2002. Knowledge spillovers in Europe: a patent citations analysis. *Scandinavian Journal of Economics*, 104, 531–545.
Metcalfe, J.S., 2002. Knowledge for growth and the growth of knowledge. *Journal of Evolutionary Economics*, 12, 3–15.
Metcalfe, J.S., 2003. Industrial growth and the theory of retardation: precursors of an adaptive evolutionary theory of economic change. *Revue Economique*, 54, 407–431.
Metcalfe, S.J., Foster, J. and Ramlogan, R., 2006. Adaptive economic growth. *Cambridge Journal of Economics*, 30, 7–32.
Morrison, A., 2008. Gatekeepers of knowledge within industrial districts: who they are, how they interact. *Regional Studies*, 42, 817–835.

Murmann, J.P. and Frenken, K., 2006. Towards a systematic framework for research on dominant designs, technological innovations, and industrial change. *Research Policy*, 35, 925–952.

Nadiri, M., 1980. Sectoral productivity slowdown. *American Economic Review*, 70, 349–352.

Nakamura, K. and Odagiri, H., 2005. R&D boundaries of the firm: an estimation of the double-hurdle model on commissioned R&D, joint R&D and licensing in Japan. *Economics of Innovation and New Technology*, 14, 583–615.

Nelson, R.R., 1959. The simple economics of basic scientific research. *Journal of Political Economy*, 67, 297–306.

Nelson, R.R., 1980. Production sets, technological knowledge, and R&D: fragile and overworked constructs for the analysis of productivity growth. *American Economic Review*, 70, 62–67.

Nelson, R.R., 1982. The role of knowledge in R&D efficiency. *Quarterly Journal of Economics*, 97, 453–470.

Nelson, R.R., 1986. Institutions supporting technical advance in industry. *American Economic Review*, 76, 186–189.

Nelson, R.R. (ed.), 1993. *National Systems of Innovation*. Oxford: Oxford University Press.

Nelson, R.R., 1995. Recent evolutionary theorizing about economic change. *Journal of Economic Literature*, 33, 48–90.

Nelson, R.R. and Winter, S., 1982. *An Evolutionary Theory of Economic Change*. Cambridge, MA: Harvard University Press.

Nelson, R.R. and Wright, G., 1992. The rise and fall of American technological leadership: the postwar era in historical perspective. *Journal of Economic Literature*, 30, 1931–1964.

Nesta, L., 2008. Knowledge and productivity in the world's largest manufacturing corporations. *Journal of Economic Behavior and Organization*, 67, 886–902.

Nesta, L. and Saviotti, P.P., 2005. Coherence of the knowledge base and the firm's innovative performance: evidence from the U.S. pharmaceutical industry. *Journal of Industrial Economics*, 53, 123–142.

Nesta, L. and Saviotti, P.P., 2006. Firm knowledge and market value in biotechnology. *Industrial and Corporate Change*, 15, 625–652.

Ngai, L.R. and Pissarides, C.A., 2007. Structural change in a multi-sector model of growth. *American Economic Review*, 97, 429–443.

Nicholas, T., 2009. Spatial diversity in invention: evidence from the early R&D labs. *Journal of Economic Geography*, 9, 1–31.

Nightingale, P., 1998. A cognitive model of innovation. *Research Policy*, 27, 689–709.

Nooteboom, B., 2000. *Learning and Innovation in Organizations and Economies*. Oxford: Oxford University Press.

OECD, 2001. *Measuring Productivity: Measurement of Aggregate and Industry-Level Productivity Growth*. Paris: OECD.

OECD, 2009. Main science and technology indicators, OECD Science, Technology and R&D Statistics (database) (accessed June 2011).

Olsson, O., 2000. Knowledge as a set in idea space: an epistemological view on growth. *Journal of Economic Growth*, 5, 253–276.

Olsson, O. and Frey, B.S., 2002. Entrepreneurship as recombinant growth. *Small Business Economics*, 19, 69–80.

Paci, R. and Usai, S., 2009. Knowledge flows across European regions. *Annals of Regional Science*, 43, 669–690.

Pasinetti, L., 1981. *Structural Change and Economic Growth: A Theoretical Essay on the Dynamics of the Wealth of Nations*. Cambridge: Cambridge University Press.

Pasinetti, L., 1993. *Structural Economic Dynamics: A Theory of the Economic Consequences of Human Learning*. Cambridge: Cambridge University Press.

Patel, P. and Soete, L., 1988. L'évaluation des effects économiques de la technologie. *STI Review*, 4, 133–183.

Patrucco, P.P., 2009. Collective knowledge production, costs and the dynamics of technological systems. *Economics of Innovation and New Technology*, 18, 295–310.

Pavitt, K., 1985. Patent statistics as indicators of innovative activities: possibilities and problems. *Scientometrics*, 7, 77–99.

Pavitt, K., 1998. Technologies, products & organisation in the innovating firm: what Adam Smith tells us and Joseph Schumpeter doesn't. *Industrial and Corporate Change*, 7, 433–457.

Perez, C., 1985. Microelectronics, long waves and world structural change: new perspectives for developing countries. *World Development*, 13, 441–463.

Perez, C., 1987. The new technologies: an integrated view, originally published in Ominami, C. (ed.), 1986. *La Tercera Revolucion Industrial: Impactos Internacionales del Actual Viraje Technologico*. Buenos Aires: Grupo Editor Latinamerican, pp. 44–89.

Perloff, H.S., Dunn, E.S., Lampard, E.E. and Muth, R.F., 1960. *Regions, Resources and Economic Growth*. Baltimore, MD: Johns Hopkins University Press.

Perroux, F., 1950. Economic space: theory and applications. *Quarterly Journal of Economics*, 64, 89–104.

Perroux, F., 1955. Note sur la notion de "pole de croissance". *Èconomie Appliquèe*, 7, 307–320.

Perroux, F., 1971. Structuralisme, modèles économiques, structures économiques, *Economie Appliquée*, 24, 3, 329–351.

Piaget, J., 1968. *Le structuralisme*. Paris: PUF.

Polanyi, M., 1958. *Personal Knowledge: Towards a Post-critical Philosophy*. Chicago: University of Chicago Press.

Polanyi, M., 1966. *The Tacit Dimension*. London: Routledge & Kegan Paul.

Pettenati, P., 1991. I Modelli di organizzazione economica del territorio, in G. Fuà (ed.), *Orientamenti per la Politica del Territorio*. Bologna: il Mulino, pp. 127–175.

Pisano, G.P., 1996. Learning before doing in the development of new process technology. *Research Policy*, 25, 1097–1119.

Ponds, R., van Oort, F. and Frenken, K., 2010. Innovation, spillovers and university–industry collaboration: an extended knowledge production function approach. *Journal of Economic Geography*, 10, 231–255.

Powell, W.W., White, D.R., Koput, K.W. and Owen-Smith, J., 2005. Network dynamics and field evolution: the growth of inter-organizational collaboration in the life sciences. *American Journal of Sociology*, 110, 1132–1205.

Price, D.S., 1976. A general theory of bibliometric and other cumulative advantage processes. *Journal of the American Society for Information Science*, 27, 292–306.

Quatraro, F., 2009a. The diffusion of regional innovation capabilities: evidence from Italian patent data. *Regional Studies*, 43, 1333–1348.

Quatraro, F., 2009b. Innovation, structural change and productivity growth: evidence from Italian regions, 1980–2003. *Cambridge Journal of Economics*, 33, 1001–1022.

Quatraro, F., 2010. Knowledge coherence, variety and productivity growth: manufacturing evidence from Italian regions. *Research Policy*, 39, 1289–1302.

Quatraro, F., 2011. ICT capital and services complementarities: the Italian evidence. *Applied Economics*, 43, 2603–2613.

Quesnay, F., 1758 [2003]. Tableau economique, in H. Hagemann, M. Landesmann and R. Scazzieri (eds), *The Economics of Structural Change*, Vol. I, *Critical Writings*, 3rd edn. Aldershot: Edward Elgar.

Ragot, X., 2003. L'économie est-elle structuraliste: un essai d'épistemologie. *Année de la Régulation*, 7, 91–111.

Ricardo, D., 1817 [2003]. On rent, in H. Hagemann, M. Landesmann and R. Scazzieri (eds), *The Economics of Structural Change*, Vol. I, *Critical Writings*. Aldershot: Edward Elgar.

Richardson, G.B., 1972. The organisation of industry. *Economic Journal*, 82, 883–896.

Romer, P., 1990. Endogenous technological change. *Journal of Political Economy*, 98, S71–S102.

Romijn, H. and Albu, M., 2002. Innovation, networking and proximity: lessons from small high technology firms in the UK. *Regional Studies*, 36, 81–86.

Rosenberg, N., 1982. *Inside the Black Box: Technology and Economics*. Cambridge: Cambridge University Press.

Rosenberg, N., 1990. Why do firms do basic research (with their own money)? *Research Policy*, 19, 2, 165–174.

Rostow, W.W., 1975. Kondratieff, Schumpeter and Kuznets: trend periods revised. *Journal of Economic History*, 35, 719–753.

Sabidussi, G., 1966. The centrality index of a graph. *Psychometrika*, 31, 581–603.

Salter, A.J. and Martin, B.R., 2001. The economic benefits of publicly funded basic research: a critical review. *Research Policy*, 30, 509–524.

Saviotti, P.P., 1988. Information, variety and entropy in technoeconomic development. *Research Policy*, 17, 89–103.

Saviotti, P.P., 2004. Considerations about the production and utilization of knowledge. *Journal of Institutional and Theoretical Economics*, 160, 100–121.

Saviotti, P.P., 2007. On the dynamics of generation and utilisation of knowledge: the local character of knowledge. *Structural Change and Economic Dynamics*, 18, 387–408.

Saviotti, P.P., 2009. Knowledge networks: structure and dynamics, in A. Pyka and A. Scharnorst (eds), *Innovation Networks: Developing an Integrated Approach*. Heidelberg: Springer Verlag.

Saviotti, P.P., 2011. Knowledge, complexity and networks, in C. Antonelli (ed.), *Handbook on the Economic Complexity of Technological Change*. Cheltenham: Edward Elgar.

Saviotti, P.P. and Catherine, D., 2008. Innovation networks in biotechnology, in H. Patzelt, T. Brenner and D. Audretsch (eds), *Handbook of Bioentrepreneurship*. Berlin: Springer.

Saviotti, P. and Frenken, K., 2008. Export variety and the economic performance of countries. *Journal of Evolutionary Economics*,18, 201–218.

Scherer, F.M., 1980. *Industrial Market Structure and Economic Performance*. Boston: Houghton Mifflin.

Schmookler, J., 1954. The level of inventive activity. *Review of Economics and Statistics*, 36, 2, 183–190.

Schmookler, J., 1962. Economic sources of inventive activity. *Journal of Economic History*, 22, 1–20.

Schmookler, J., 1966. Patent statistics, in *Invention and Economic Growth*. Cambridge, MA: Harvard University Press, pp. 18–56.

Schumpeter, J.A., 1911 [1934]. *The Theory of Economic Development.* Oxford: Galaxy Books.
Schumpeter, J., 1912. *The Theory of Economic Development.* Cambridge, MA: Harvard University Press.
Schumpeter, J.A., 1928. The instability of capitalism. *Economic Journal*, 38, 361–386.
Schumpeter, J.A., 1936. The general theory of employment, interest and money, by John Maynard Keynes. *Journal of the American Statistical Association*, 31, 791–795.
Schumpeter, J.A., 1939. *Business Cycles: A Theoretical, Historical and Statistical Analysis of the Capitalist Process.* New York and London: McGraw Hill.
Schumpeter, J.A., 1942. *Capitalism, Socialism and Democracy.* London: Unwin.
Schumpeter, J.A., 1946. John Maynard Keynes 1883–1946. *American Economic Review*, 36, 495–518.
Schumpeter, J.A., 1947. The creative response in economic history. *Journal of Economic History*, 7, 149–159.
Scott, J., 2000. *Social Network Analysis: A Handbook.* London: Sage.
Shannon, C.E., 1948. A mathematical theory of communication. *Bell System Technical Journal*, 27, 379–423.
Silva, E.G. and Teixeira, A.A.C., 2008. Surveying structural change: seminal contributions and a bibliometric accounts. *Structural Change and Economic Dynamics.*
Simon, H., 1962. The architecture of complexity: hierarchic systems. *Proceedings of the American Philosophical Society*, 106, 467–482.
Simon, H., 1966. *The Sciences of the Artificial.* Cambridge, MA: MIT Press.
Simon, H., 1969. *Models of Discovery.* Boston: D. Reidel.
Simon, H., 1973. The organization of complex systems, in H. Pattee (ed.), *Hierarchy Theory: The Challenge of Complex Systems.* New York: George Braziller.
Simon, H.A., 1982. *Models of Bounded Rationality.* Cambridge, MA: MIT Press.
Skvoretz, J., 1991. Theoretical and methodological models of networks and relations. *Social Networks*, 13, 275–300.
Smith, A., 1776 [1986]. *The Wealth of Nations: Books I–III.* New York: Penguin.
Soete, L., 1987. The impact of technological innovation on international trade patterns: the evidence reconsidered. *Research Policy*, 16, 101–130.
Solow, R.M., 1956. A contribution to the theory of economic growth. *Quarterly Journal of Economics*, 70, 65–94.
Solow, R.M., 1957. Technical change and the aggregate production function. *Review of Economics and Statistics*, 39, 312–320.
Sorenson, O., Rivkin, J.W. and Fleeming, L., 2006. Complexity, network and knowledge flows. *Research Policy*, 35, 994–1017.
Steinmuller, W.E., 2002. Knowledge-based economies and information and communication technologies. *International Social Science Journal*, 54, 141–153.
Stirling, A., 2007. A general framework for analysing diversity in science, technology and society. *Journal of the Royal Society Interface*, 4, 707–719.
Syrquin, M., 1988. Patterns of structural change, in H.B. Chenery and T.N. Srinivasan (eds), *Handbook of Development Economics.* Amsterdam: North-Holland.
Syrquin, M., 2010. Kuznets and Pasinetti on the study of structural transformation: never the twain shall meet? *Structural Change and Economic Dynamics*, 21, 248–257.
Syrquin, M. and Chenery, H.B, 1989. Three decades of industrialization. *World Bank Economic Review*, 3, 145–181.
Teece, D.J., 1988. Technological change and the nature of the firm, in G. Dosi, C. Freeman, R. Nelson, G. Silverberg and L. Soete (eds), *Technical Change and Economic Theory.* New York: Pinter Publishers, pp. 256–281.

Teece, D.J., Rumelt, R., Dosi, G. and Winter, S., 1994. Understanding corporate coherence: theory and evidence. *Journal of Economic Behavior and Organization*, 23, 1–30.

Theil, H., 1967. *Economics and Information Theory*. Amsterdam: North-Holland.

Théret, B., 2003. Institutionnalismes et structuralismes: oppositions, substitutions ou affinités électives? *Cahiers d'économie politique*, 44, 51–78.

Thomas, M.J., 1975. Growth pole theory, technological change and regional economic growth. *Papers in Regional Science*, 34, 3–25.

Timmer, M.P. and van Ark, B., 2005. Does information and communication technology drive EU–US growth differentials? *Oxford Economic Papers*, 57, 693–716.

Tinbergen, 1952. De quelques problèmes posés par le concept de structure économique. *Revue Economie appliquée*, 27–46.

Trajtenberg, M., 1990. A penny for your quotes: patent citations and the value of innovations. *Rand Journal of Economics*, 21, 172–187.

Tushman, M.L. and Anderson, P., 1986. Technological discontinuities and organizational environments. *Administrative Science Quarterly*, 31, 439–465.

Ulku, H., 2007. R&D, innovation, and growth: evidence from four manufacturing sectors in OECD countries. *Oxford Economic Papers*, 59, 513–535.

Usher, A., 1954. *A History of Mechanical Invention*. Cambridge, MA: Dover.

Van den Ende, J. and Dolfsma, W., 2005. Technology-push, demand-pull and the shaping of technological paradigms: patterns in the development of computing technology. *Journal of Evolutionary Economics*, 15, 83–99.

Verspagen, B., 1995a. R&D and productivity: a broad cross-section cross-country look. *Journal of Productivity Analysis*, 6, 117–135.

Verspagen, B., 1995b. The dynamics of technology, trade and growth. *Journal of Economic Literature*, 33, 1515–1515.

Verspagen, B., 1997. Estimating international technology spillovers using technology flow matrices. *Review of World Economics (Weltwirtschaftliches Archiv)*, 133, 2, 226–248.

Von Hippel, E., 1988. *The Sources of Innovation*. Oxford: Oxford University Press.

von Hippel, E., 1994. Sticky information and the locus of problem solving: implications for innovation. *Management Science*, 40, 429–439.

von Hippel, E., 2005. *Democratizing Innovation*. Cambridge, MA: MIT Press.

Wasserman, S. and Faust, K., 2007. *Social Network Analysis: Method and Applications*. Cambridge: Cambridge University Press.

Weiller, J., 1935. *Structure économique et commerce extérieur*. Paris: Sirey.

Weiller, J., 1952. Les analyses de structure et le déséquilibre mondial. *Revue Économique*, 3, 856–869.

Weitzman, M.L., 1996. Hybridizing growth theory. *American Economic Review*, 86, 207–212.

Weitzman, M.L., 1998. Recombinant growth. *Quarterly Journal of Economics*, 113, 331–360.

Wolff, E.N., 2003. What's behind the rise in profitability in the US in the 1980s and the 1990s? *Cambridge Journal of Economics*, 27, 479–499.

Yayavaram, S. and Ahuja, G., 2008. Technological search and decomposability in knowledge structures: impact on invention utility and knowledge base malleability. *Administrative Science Quarterly*, 53, 333–362.

Young, A., 1928. Increasing returns and economic progress. *Economic Journal*, 38, 527–542.

Zucker, L.G., Darby, M.R. and Armstrong, J.S., 1998. Geographically localized knowledge: spillovers or markets? *Economic Inquiry*, 36, 65–86.

Index

Page numbers in *italics* denote tables, those in **bold** denote figures.

Taylor & Francis

eBooks

FOR LIBRARIES

ORDER YOUR FREE 30 DAY INSTITUTIONAL TRIAL TODAY!

Over 23,000 eBook titles in the Humanities, Social Sciences, STM and Law from some of the world's leading imprints.

Choose from a range of subject packages or create your own!

Benefits for you

- Free MARC records
- COUNTER-compliant usage statistics
- Flexible purchase and pricing options

Benefits for your user

- Off-site, anytime access via Athens or referring URL
- Print or copy pages or chapters
- Full content search
- Bookmark, highlight and annotate text
- Access to thousands of pages of quality research at the click of a button

For more information, pricing enquiries or to order a free trial, contact your local online sales team.

UK and Rest of World: **online.sales@tandf.co.uk**

US, Canada and Latin America:
e-reference@taylorandfrancis.com

www.ebooksubscriptions.com

ALPSP Award for BEST eBOOK PUBLISHER 2009 Finalist

Taylor & Francis eBooks
Taylor & Francis Group

A flexible and dynamic resource for teaching, learning and research.